AF576032

Biomarkers Used for the Diagnosis of Diseases

Biomarkers Used for the Diagnosis of Diseases

Editors

Satish Balasaheb Nimse
Min Park

MDPI • Basel • Beijing • Wuhan • Barcelona • Belgrade • Manchester • Tokyo • Cluj • Tianjin

Editors
Satish Balasaheb Nimse
Hallym University
Korea

Min Park
Hallym University
Korea

Editorial Office
MDPI
St. Alban-Anlage 66
4052 Basel, Switzerland

This is a reprint of articles from the Special Issue published online in the open access journal *Biosensors* (ISSN 2079-6374) (available at: https://www.mdpi.com/journal/biosensors/special_issues/biomarkers_diagnosis_diseases).

For citation purposes, cite each article independently as indicated on the article page online and as indicated below:

LastName, A.A.; LastName, B.B.; LastName, C.C. Article Title. *Journal Name* **Year**, *Volume Number*, Page Range.

ISBN 978-3-0365-3569-2 (Hbk)
ISBN 978-3-0365-3570-8 (PDF)

Cover image courtesy of Satish Balasaheb Nimse and Min Park.

Contents

About the Editors

Satish Balasaheb Nimse earned a Bachelor of Pharmaceutical Sciences at the University of Pune, Maharashtra, India, and a Master of Medicinal Chemistry at Biju Pattanaik University Technology, Orissa, India. He obtained a Ph.D. in organic chemistry from the Department of Chemistry, Hallym University, South Korea. He joined as an assistant professor in the Department of Chemistry, Hallym University, South Korea, and has served in several positions to date. His research interests include medicinal chemistry, bioorganic chemistry, molecular diagnostics, and the applications of supramolecules for the fabrication of diagnostic DNA chips and drug discovery. He has published over 60 scientific articles in reputed international journals.

Min Park is a tenured associate professor in Materials Science and Engineering at Hallym University, Chuncheon, Korea. He received his B.S. in Ceramic Engineering from Yonsei University, Seoul, Korea. He received his Ph.D. degree in Materials Science and Engineering from Yonsei University, Seoul, Korea in 2013. He has worked at Westfälische Wilhelms-University, Münster, Germany and Korea Institute of Science and Technology, Seoul, Korea as a post-doctoral researcher during 2013–2016. He joined Hallym University as an assistant professor in 2016, and then he was promoted to associate professor with tenure in 2022. His research interests include the development of bioassays, biosensors, medical diagnosis, surface display of active enzymes and affinity proteins, biochip technology, and molecular diagnostics.

Preface to "Biomarkers Used for the Diagnosis of Diseases"

Recent advances in chemical technology and biotechnology have rapidly changed the landscape of molecular diagnostics. Research on the discovery of novel biomarkers that allow for the rapid diagnosis of clinical ailments, where early detection is a key to survival, has seen exponential growth in recent years. Interestingly, the discovery of novel biomarkers is supported by the rapid growth in the development of platform technologies that allow for the highly sensitive and specific detection of biomarkers. Nucleic acid biomarkers (micro RNA, cDNA, and ctDNA), protein biomarkers (cTnT, PSA, CRP, and CYFRA 21-1), and small biomolecules (cysteine, homocysteine, and glutathione) are some of the biomarkers that are always under the scrutiny of the scientific community, who are working effortlessly to make disease diagnosis simple, rapid, and highly applicable in resource-limited settings.

The detection and quantification with high precision of nucleic acid biomarkers and protein biomarkers in resource-limited settings will be key to the early diagnosis of diseases and for monitoring the effects of treatments. As there is an enormous demand for high-quality biomarker detection platforms that are robust and highly applicable in resource-limited settings, this book is devoted to exploring methods for detection and quantification of biomarkers, focusing on the recent advances in this field.

Satish Balasaheb Nimse and Min Park
Editors

 *biosensors* 

Article

Development of a Novel Benzimidazole-Based Probe and Portable Fluorimeter for the Detection of Cysteine in Human Urine

Gyu Seong Yeom [1,†], In-ho Song [1,†], Shrikant Dashrath Warkad [2], Pramod B. Shinde [3], Taewoon Kim [4], Seong-min Park [1] and Satish Balasaheb Nimse [1,*]

1 Institute of Applied Chemistry and Department of Chemistry, Hallym University, Chuncheon 24252, Korea; m21019@hallym.ac.kr (G.S.Y.); songin-ho@hallym.ac.kr (I.-h.S.); psm3561@naver.com (S.-m.P.)
2 Biometrix Technology, Inc., 2-2 Bio Venture Plaza 56, Chuncheon 24232, Korea; shrikant@bmtchip.com
3 Natural Products & Green Chemistry Division, CSIR—Central Salt and Marine Chemicals Research Institute (CSIR-CSMCRI), Council of Scientific and Industrial Research (CSIR), Bhavnagar 364002, Gujarat, India; pramodshinde@csmcri.res.in
4 School of Software, Hallym University, Chuncheon 24252, Korea; taewoon@hallym.ac.kr
* Correspondence: satish_nimse@hallym.ac.kr
† These authors contributed equally to this work.

Abstract: The measurement of cysteine in human urine and live cells is crucial for evaluating biological metabolism, monitoring and maintaining the immune system, preventing tissue/DNA damage caused by free radicals, preventing autoimmune diseases, and diagnosing disorders such as cystinuria and cancer. A method that uses a fluorescence turn-on probe and a portable fluorescence spectrometer device are crucial for highly sensitive, simple, rapid, and inexpensive cysteine detection. Herein, we present the synthesis and application of a benzimidazole-based fluorescent probe (**ABIA**) along with the design and development of a portable fluorescence spectrometer device (**CysDDev**) for detecting cysteine in simulated human urine. **ABIA** showed excellent selectivity and sensitivity in detecting cysteine over homocysteine, glutathione, and other amino acids with the response time of 1 min and demonstrated a detection limit of 16.3 nM using the developed **CysDDev**. Further, **ABIA** also demonstrated its utility in detecting intracellular cysteine, making it an excellent probe for bio-imaging assay.

Keywords: cysteine; biothiols; cystinuria; portable; fluorimeter; bio-imaging; cancer

Citation: Yeom, G.S.; Song, I.-h.; Warkad, S.D.; Shinde, P.B.; Kim, T.; Park, S.-m.; Nimse, S.B. Development of a Novel Benzimidazole-Based Probe and Portable Fluorimeter for the Detection of Cysteine in Human Urine. *Biosensors* **2021**, *11*, 420. https://doi.org/10.3390/bios11110420

Received: 16 September 2021
Accepted: 25 October 2021
Published: 26 October 2021

Publisher's Note: MDPI stays neutral with regard to jurisdictional claims in published maps and institutional affiliations.

1. Introduction

Biothiols such as cysteine (Cys), homocysteine (Hcy), and glutathione (GSH) play a vital role in numerous biological reactions [1–6]. It is important to note that among other biothiols, Cys deficiency exhibits several disorders, including edema, liver damage, muscle and fat loss, narcolepsy, skin damage, and weakness [7–9]. On the other hand, elevated cysteine levels are often correlated with cystinuria, a metabolic disorder characterized by the urinary loss of amino acids including Cys, arginine, ornithine, and lysine [10]. Therefore, measuring Cys in human urine and live cells is crucial for evaluating biological metabolism, monitoring and maintaining the immune system, preventing tissue/DNA damage caused by free radicals, preventing autoimmune diseases, and diagnosing disorders such as cystinuria and cancer.

There have been tremendous advances in developing methods for detecting biothiols in serum and live cells [11–14]. The development of fluorescent probes for detecting Cys has attracted significant attention from the broader scientific community [15–18]. The simplicity, high sensitivity and selectivity, and suitability for noninvasive and real-time detection are the significant advantages of methods based on fluorescent probes [19–22]. Though there are numerous reports on mechanistically different fluorescent probes for detecting biothiols [23–26], only a few have been applied to simultaneously detect Cys in human urine and live-cell imaging [27–29]. Therefore, using a fluorescent probe that can detect Cys in human urine and live cells can result in an excellent method for diagnosing cystinuria and Cys-related disorders [30].

There have been several reports on the colorimetric detection of Cys using paper-based devices [31,32]. However, proteins and large quantities of amino acids can interfere in color development reactions. The other disadvantage of paper-based devices is a higher limit of detection (LOD) [33,34]. Hence, paper-based devices can only provide qualitative information on urinary cysteine. Further, the fluorescence-based detection of Cys in biological fluids is advantageous over other methods due to its simple operating procedure, high sensitivity and selectivity for Cys in the presence of other proteins and amino acids, lower LOD, and the possibility of developing a portable device applicable in point-of-care settings.

Herein, we report on developing and applying a Cys specific fluorescent probe and the portable device for detecting cysteine in urine. The probe **ABIA** (2-(anthracen-9-yl)-5-methyl-1H-benzo[d]imidazole-acrylamide), an organic dye, was prepared in a two-step reaction and was evaluated as the target probe for biothiols. **ABIA** demonstrated excellent selectivity for Cys over other biothiols and amino acids, with a response time of a mere 1 min. The application of **ABIA** in live-cell imaging was evaluated by inverted fluorescence microscopy in a lung cancer cell line (A549), proving that **ABIA** as a suitable tool for tracking Cys in vitro. Its applicability in detecting Cys in live cells and simulated urine prompted us to develop a portable fluorimeter. The design and development of Cys detection device (**CYsDDev**) is also elaborated upon here. **ABIA** and **CysDDev** allow the detection of Cys in simulated urine, with a LOD of 16.3 nM.

2. Materials and Methods

2.1. Materials and Instrumentation

Reagents and required chemicals were procured either from Sigma-Aldrich (Seoul, Korea) or TCI (Seoul, Korea) and were used as received unless stated otherwise. The simulated urine (product code, 84 0679) was obtained from Biozoa Biological Supply Company (Seoul, Korea). The reactions were conducted under an inert atmosphere (argon gas). Completion of the reaction was monitored using thin-layer chromatography (TLC) by visualizing the plates under UV light (254 nm). DMSO-*d6* was used as a solvent to characterize synthesized compounds by ^{1}H- and ^{13}C-NMR spectroscopy on a Jeol FT-NMR spectrometer (400 MHz; JEOL, Tokyo, Japan). The chemical shifts (δ) and the coupling constants (J) were reported as ppm and Hz, respectively. The Shimadzu UV-24500 (Shimadzu, Tokyo, Japan) spectrometer and Agilent Cary Eclipse fluorescence spectrophotometer (Agilent Technologies, Santa Clara, CA, USA) were used to record UV-visible spectra and fluorescence emission spectra, respectively. A JMS-700 MStation Mass Spectrometer (JEOL, Tokyo, Japan), a microplate Reader Spectramax Plus 384 (Molecular Devices, San Jose, CA, USA), and a confocal laser scanning microscope (Carl Zeiss LSM710, Osnabrück, Germany) were also used in the present study. The material for 3D printing and the electronic components required for the development of **CysDDev** were procured from Devicemart (Seoul, Korea) and Mouser Electronics (Seoul, Korea). A CUBICON Single Plus 320C (Jungwon, Korea) 3D printer was used to fabricate various parts and the body of **CysDDev**.

2.2. Synthesis of Compound **3** (2-(Anthracen-9-yl)-5-methyl-1H-benzo[d]imidazole)

A solution of anthracene-9-carbaldehyde (0.84 g, 4.07 mmol) in ethanol (10 mL) was slowly added dropwise to a solution of 4-methylbenzene-1,2-diamine (0.5 g, 4.09 mmol) in ethanol (20 mL) with a dropping funnel for 2 h. Then, the reaction mixture was continually stirred under air at room temperature until the completion of the reaction. The progress of the reaction was monitored by TLC using an ethyl acetate:hexanes (3:7) mixture. The precipitated product was filtered. The filter cake was washed with three aliquots of 10 mL ethanol followed by vacuum drying to obtain the final product: beige-colored powder. Yield: 1.01 g (80%). ^{1}H-NMR (400 MHz, DMSO-d_6) δ 12.88 (s, 1H), 8.84 (s, 1H), 8.21 (d, J = 8.4 Hz, 2H), 7.69 (d, J = 8.7 Hz, 2H), 7.61–7.55 (m, 3H), 7.55–7.48 (m, 3H), 7.39 (s, 1H), 2.50 (s, 3H). ^{13}C-NMR (100 MHz, DMSO-d_6) δ 130.68, 130.58, 128.74, 128.50, 126.80, 126.02, 125.65, 21.34; HR-EIMS$^+$: *m/z* = 308.1311 $(M)^+$ (Calcd for $C_{22}H_{16}N_2$, 308.13).

2.3. Synthesis of Probe **ABIA** (1-(2-(Anthracen-9-yl)-5-methyl-1H-benzo[d]imidazol-1-yl)prop-2-en-1-one)

A solution of compound **3** (0.1 g, 0.32 mmol), triethylamine (0.18 mL, 1.29 mmol), and DMAP (0.004 g, 0.033 mmol) in anhydrous tetrahydrofuran (25 mL) was stirred at 0 °C under argon gas for 30 min. Then, prop-2-enoyl chloride (0.104 mL, 1.29 mmol) was slowly added to the reaction mixture and continued to be stirred until the completion of the reaction. The reaction progress was monitored by TLC using an ethyl acetate:hexanes (3:7) mixture. After the completion of the reaction (1 h), 100 mL of CH_2Cl_2 was added to the reaction mixture. Then, the reaction mixture was washed three times with the 150 mL aliquots of saturated Na_2CO_3 followed by a single wash of 150 mL brine. The separated organic layer was then dried over anhydrous $MgSO_4$, and the solvent was evaporated under reduced pressure to obtain the crude product, which was then purified by column chromatography (SiO_2, ethyl acetate/hexane = 1/20, *v*/*v*). The final compound was obtained as a light yellow solid. Yield 82 mg (74%). ^{1}H-NMR (400 MHz, DMSO-d_6) δ 8.92 (s, 1H), 8.23 (d, J = 8.0 Hz, 2H), 8.13–7.96 (m, 1H), 7.85–7.69 (m, 1H), 7.59 (dd, J = 8.3, 4.4 Hz, 2H), 7.54 (d, J = 3.2 Hz, 4H), 7.38 (t, J = 9.8 Hz, 1H), 5.99 (dd, J = 16.9, 3.6 Hz, 1H), 5.71 (ddd, J = 16.9, 10.5, 6.5 Hz, 1H), 5.27 (d, J = 10.9 Hz, 1H), 2.54 (d, J = 14.1 Hz, 3H). ^{13}C NMR (100 MHz, DMSO-d_6) δ 164.38, 164.20, 149.45, 148.87, 143.21, 141.05, 135.18, 134.53, 133.39, 132.52, 131.19, 130.71 (A), 130.68 (B), 130.53, 130.01, 128.82, 128.37, 128.25, 127.81, 126.65, 126.25, 125.86, 124.75 (A), 124.62 (B), 119.90 (A), 119.63 (B), 115.16 (A), 114.89 (B), 21.60 (A), 21.08 (B). ((A), (B) represents peak separation due to tautomerization of **ABIA**, details are shown in Supporting information Figure S7 and Table S1). HR-EIMS$^+$: *m/z* = 362.1418 $(M)^+$ (Calcd for $C_{25}H_{18}N_2O$, 362.14).

2.4. UV–Vis and Fluorescence Spectroscopy

The double-distilled water and spectroscopic grade DMSO were used to prepare the stock and working solutions. The working solution of **ABIA** (10 μM) was prepared by diluting the stock solution of **ABIA** (1 mM, in DMSO) with a mixture of 90% DMSO:0.01 M HEPES buffer. The working solutions (1 mM) of amino acids were prepared by diluting the stock solutions (10 mM, in double-distilled) with 0.01 M HEPES buffer. The UV-visible absorption and emission spectra of **ABIA** (10 μM) dissolved in 90% DMSO:0.01 M HEPES buffer were recorded at room temperature (298 K) by adding the solution of various analytes such as glycine (Gly), alanine (Ala), serine (Ser), proline (Pro), valine (Val), threonine (Thr), isoleucine (Ile), leucine (Leu), asparagine (Asp), aspartic acid (Asn), glutamine (Gln), lysine (Lys), glutamic acid (Glu), methionine (Met), histidine (His), phenylalanine (Phe), arginine (Arg), tyrosine (Tyr), tryptophan (Typ), homocysteine (Hcy), glutathione (GSH), cysteine (Cys), bovine serum albumin (BSA), and other biologically relevant ions, including Cl^-, Br^-, NO_3^-, AcO^-, SO_4^{2-}, PO_4^{3-}, Na^+, Cs^+, Ca^{2+}, and Cu^{2+} to examine the selectivity of **ABIA**.

For sensitivity study, the titration experiments were accomplished through a stepwise addition of 1.15 equivalents of Cys (1 mM) to a solution of **ABIA** (10 μM). The absorbance intensity and emission intensity were recorded in the range of 200–600 nm and 350–650 nm, respectively, alongside a reagent blank. The effect of reaction time on the Cys detection was determined by tracing the fluorescence intensity of **ABIA** (10 μM) in the presence of five equivalent Cys (50 μM) for 0 to 60 min. The response of **ABIA** for reaction with Cys obeyed pseudo-first-order kinetics. Therefore, the rate constant was calculated according to the following equation (Equation (1)) [35]:

$$\ln((F_{max} - F)/F_{max}) = -kt \quad (1)$$

where F is the fluorescence intensity at time t, F_{max} is the fluorescence intensity after the reaction, and k is the rate constant. The fluorescence intensity was recorded at $\lambda_{ex}/\lambda_{em}$ = 368/470 nm alongside a reagent blank with the excitation and emission slits set to 5.0 nm.

2.5. Fabrication of *CysDDev*

The 3D models of various parts and the main case were created in Autodesk Fusion 360 software. The device's overall size was 130 × 120 × 50 mm (W × D × H). **CysDDev** was assembled with several parts, including the cuvette holder, mounting holes for the LED and sensor, lid for the LED and sensor, the main case, and a top cap lid for the cuvette holder. The cuvette holder could hold a plastic or quartz cuvette with the dimensions of 12.5 × 12.5 × 45 mm dimension. The main body had a hole connecting the device to the computer. When completely assembled and when the top cap lid was mounted, there was no significant light entering the measuring device. Apart from the top cap lid for the cuvette holder, there were no moving components in the device, making it easy to fabricate and assemble. The **CysDDev** prototype was optimized for fused deposition modeling (FDM) 3D printing. All three parts of the design were printed separately with black thermoplastic filament to lower the reflectivity of the internal surfaces. The electronics parts of the **CYSDDev** consisted of 9 V battery, a sensor (Hamamatsu micro-spectrometer C12666MA, Japan) with the detection range λ = 340–850 nm), a UV LED (15 mA, λ = 370 nm), an Arduino Uno R3 microcontroller for device control and data processing, and a computer connector cable. The LED drew power from a 9 V battery, and the Arduino microcontroller received the energy from the connected computer. The measured fluorescence profile of a sample by **CysDDev** was automatically converted into a Microsoft Excel file (.csv) consisting of the information on the wavelength (nm) and respective fluorescence intensity (a.u.).

2.6. pH Effect on Cys Detection by *ABIA*

The effect of pH (pH = 3–10) on the Cys detection using **ABIA** was examined by fluorescence spectroscopy. The tetrabutylammonium hydroxide and perchloric acid were used to adjust the pH of simulated human urine.

2.7. Detection of Cys Using *ABIA* and *CysDDev* in Simulated Human Urine and Real Urine Sample

Usually, urine pH is slightly acidic, with typical values of 6.0 to 7.5. However, the range of urine pH is 4.5 to 8.0. Cys detection using **ABIA** and **CysDDev** was performed using simulated urine with a pH of 6.5. A standard curve was obtained by diluting a stock solution of Cys in analyte-free simulated human urine (0–150 μM). Each sample was serially diluted with the analyte-free urine to obtain solutions with concentrations in the detection range. The mean of ten fluorescence signal measurement values (SD in the range of 1.5–9.8%) for each calibration point was used to construct the standard curve. The LOD was estimated by applying the IUPAC recommended equation, LOD = 3σ/slope [36]. Where σ is the standard deviation of (n = 10) blank samples and the slope for the calibration curves.

A real urine sample obtained from a healthy volunteer was used for the spiking test and to validate the applicability of the **ABIA** and **CysDDev** in clinical samples. An amount of 5 μL urine from a healthy volunteer spiked with various Cys concentrations was added to the **ABIA** solution, and the fluorescence intensities were measured. Then, the percent recovery was measured using the standard curve.

2.8. Cell Culture, Cytotoxicity Assay, and Bio-Imaging

The cytotoxicity assay (MTT (3-(4,5-dimethylthiazol-2-yl)-2,5-diphenyltetrazolium bromide) assay) of **ABIA**, Cys, and Cys combined with the probe **ABIA** was performed using A549 cells (adenocarcinomic human alveolar basal epithelial cells), as reported earlier [37,38]. In brief, about 7000 A549 cells per well were seeded in 96-well plates and were allowed to incubate for 24 h. Then, the solution in the wells was replaced by media containing **ABIA**, Cys, and **ABIA** with Cys (1, 10, 25, and 50 μM) and was incubated for another 24 h. The dimethyl sulfoxide (DMSO) was used as a control. About 200 μL of a media containing MTT solution was added to each well and was allowed to incubate for four hours at 37 °C. Finally, the absorbance was recorded at 570 nm to determine the cell cytotoxicity. The experiment was repeated three times, and the results were presented with mean and standard deviation.

The bio-imaging experiments were performed by seeding the A549 cells on glass-bottom culture dishes and incubating them for 48 h. For control experiments, cells were only treated with the culture medium and were observed under the confocal microscope after being washed three times with 0.1 M PBS (0.5 mL). For treatment, cells were treated with 500 μM of Cys and 25 μM of **ABIA** separately and were allowed to incubate for 1 h at 37 °C and 5% CO_2 and were then washed three times with 0.1 M PBS (0.5 mL) before observation. For the bio-imaging of Cys by **ABIA**, cells were incubated with 500 μM of Cys for 1 h followed by replacing the media containing 25 μM of **ABIA** and incubating the cells for another 1 h. Then, the cells were washed three times with 0.1 M PBS (0.5 mL) before observation. For another control experiment, cells were treated with 1 mM N-ethylmaleimide (NEM, a biothiol scavenger) for 30 min at 37 °C followed by washing three times with 0.1 M PBS (0.5 mL). Then, the cells were incubated in the presence of 25 μM of **ABIA** for 1 h followed being washed three times with 0.1 M PBS (0.5 mL) before observation. In a similar experiment, cells pre-tread with NEM (1 mM), as mentioned above, were treated with Cys (500 μM, for 1 h) followed being washed with 0.1 M PBS (0.5 mL × 3) and then incubated in the presence of **ABIA** (25 μM, 1 h) and finally being washed three times 0.1 M PBS (0.5 mL) before observation. Finally, the PBS (1 mL) was added to each glass-bottom culture dish, and then the fluorescence images were recorded using a confocal microscope (Carl Zeiss LSM710).

3. Results

3.1. Synthesis and Characterization of the Probe ABIA

The probe **ABIA** was synthesized in a two-step reaction as depicted in Scheme 1.

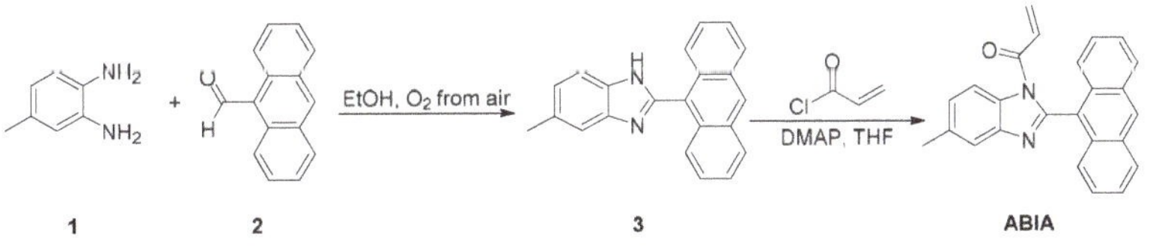

Scheme 1. Scheme for the synthesis of the probe **ABIA**.

In the first step, a simple yet efficient method for the synthesis of compound **3** was developed. The reaction of 4-methylbenzene-1,2-diamine (4.0 mmol) and anthracene-9-carbaldehyde (4.0 mmol) in ethanol at room temperature in the presence of O_2 from the air and visible light allowed us to afford compound **3** with 80% yield (see the Supplemental Information Figures S1–S3). To the best of our knowledge, this is a first report on the synthesis of compound **3**. The reported synthetic procedures to obtain benzimidazole derivatives are different than the one employed here [39,40]. The previously reported methods used stringent reaction conditions, including dimethyl sulfoxide as a solvent, the temperature of 110 °C, and I_2 as a catalyst. It is important to note that the procedure employed here uses mild reaction conditions, including ethanol as a solvent and 25 °C without using the additional catalyst. The method presented here allowed the synthesis of compound **3** with comparable yield.

In a second step, compound **3** was reacted with the prop-2-enoyl chloride (1.29 mmol) in the presence of triethylamine (1.29 mmol) and DMAP (0.033 mmol) to afford the final compound **ABIA** (see the Supplemental Information Figures S4–S6) in comparable yield.

The ^{1}H NMR and ^{13}C NMR spectra of **ABIA** demonstrate the 5(6)-methyl tautomerism related to the N-substituted benzimidazoles (see the Supplemental Information Figure S7 and Table S1). According to previous reports, the tautomerism in benzimidazole derivatives is a well-known phenomenon, and in most cases, the separation of tautomers by column chromatography is not feasible [41,42]. Hence, we used the tautomeric mixture of **ABIA** in further experiments. The photophysical properties of compound **3** (10 μM) and probe **ABIA** (10 μM) in 90% DMSO:0.01 M HEPES buffer solution did not indicate a significant effect of tautomerism in **ABIA**. As shown in Figure 1, the optical absorbance spectrum of compound **3** shows λ_{ex} at 368 nm and another two peaks at 353 nm and 388 nm. The emission spectra of compound **3** show λ_{em} at 475 nm (λ_{ex} = 368 nm), indicating Stoke's shift of 107 nm. Similarly, the probe **ABIA** demonstrates three distinct peaks (356, 372, and 392 nm) with the absorption maxima at 372 nm.

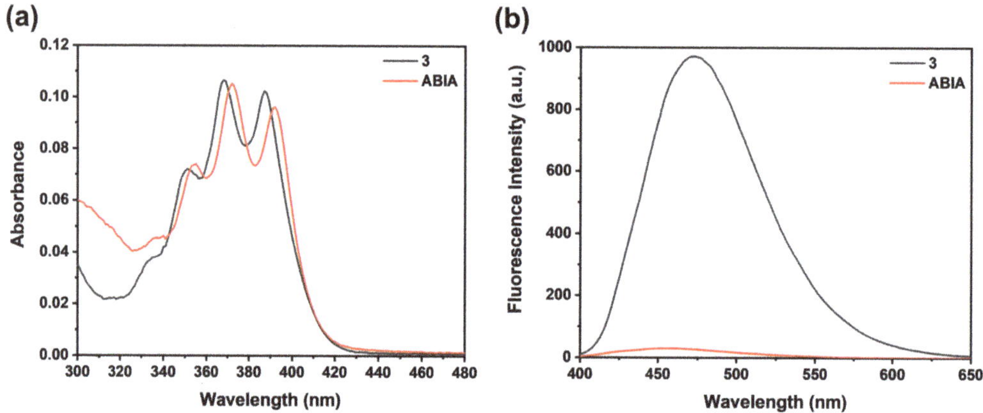

Figure 1. (**a**) UV-vis and (**b**) fluorescence emission spectra of compound **3** (10 μM; λ_{ex} = 368 nm) and the probe **ABIA** (10 μM) in 90% DMSO:0.01 M HEPES buffer solution (λ_{ex} = 372 nm).

Interestingly, **ABIA** showed fluorescence (λ_{em} = 453 nm) that was almost 30-fold lower than compound **3**, indicating that the intramolecular charge transfer induced fluorescence quenching due to the N1-substituted acrylate group. The acrylate group has a strong electron-withdrawing ability that enhances the fluorescence turn-off effect in **ABIA**. These results indicate that reaction-induced removal of the acrylate group in **ABIA** can result in compound **3** with the fluorescence turn-on effect. Hence, the spectral characteristics of compound **3** were studied using various ratios (10, 50, 70, and 90%) of the DMSO:0.01 M HEPES buffer solution. As shown in Figure S8 and Table S2, compound **3** shows aggregation-induced fluorescence quenching with the increasing water percentage. It is important to note that 10% DMSO DMSO:0.01 M HEPES buffer solution completely dissolved 10 μM of compound **3**. However, compound **3** showed maximum fluorescence in 90% DMSO:0.01 M HEPES buffer solution. Hence, other fluorimetric experiments were conducted in 90% DMSO: 0.01 M HEPES buffer solution (λ_{ex} = 368 nm).

3.2. *Determination of Selectivity of the Probe **ABIA** as a Chemosensor for Cys*

The selectivity of the probe **ABIA** for Cys detection was investigated using fluorescence spectroscopy. The fluorescence spectra of **ABIA** (10 μM, in 90% DMSO:0.01 M HEPES buffer solution) were recorded immediately (1 min) after adding the five equivalents of various amino acids, BSA, anions (Cl^-, Br^-, NO_3^-, AcO^-, SO_4^{2-}, PO_4^{3-}), and cations (Na^+, Cs^+, Ca^{2+}, Cu^{2+}) to the solution containing **ABIA**. As shown in Figure 2a, **ABIA** showed negligible fluorescence. However, the fluorescence intensity of **ABIA** increased significantly in the presence of Cys (λ_{em} = 455 nm) compared to the other analytes. Among the mercapto group-containing amino acids, **ABIA** showed 6-fold and 15-fold higher selectivity for Hcy (λ_{em} = 460 nm) and GSH (λ_{em} = 475 nm), respectively. In contrast, the selectivity for Cys was 50-fold higher than other amino acids. Further, **ABIA** did not show any interaction with biologically relevant ions such as Cl^-, Br^-, NO_3^-, AcO^-, SO_4^{2-}, PO_4^{3-}, Na^+, Cs^+, Ca^{2+}, and Cu^{2+}. As shown in Figure 2b, in competition experiments conducted in the presence of an excess of (five equivalents) other amino acids, including Hcy and GSH, **ABIA** did not show any interference in the detecting Cys. However, among the tested biologically relevant ions, only Cu^{2+} affected the detection of Cys at tested equimolar concentrations (50 µM). Cys forms a coordination complex with Cu^{2+} ions using the mercapto and carboxylate functional groups [43]. Therefore, the complexation of Cys with Cu^{2+} inhibits its nucleophilic addition ability to the acrylate group of **ABIA**. However, the amount of Cu^{2+} (180 nM) [44] is negligible compared to Cys in healthy human urine (120 µM) and in the urine of cystinuria patients (2000 µM) [45,46]. Thus, Cu^{2+} would not interfere in Cys detection using the probe **ABIA** in the actual urine sample. These results indicate that **ABIA** is a highly selective fluorescence turn-on probe for Cys, and it can be applied to detect intracellular and urinary Cys.

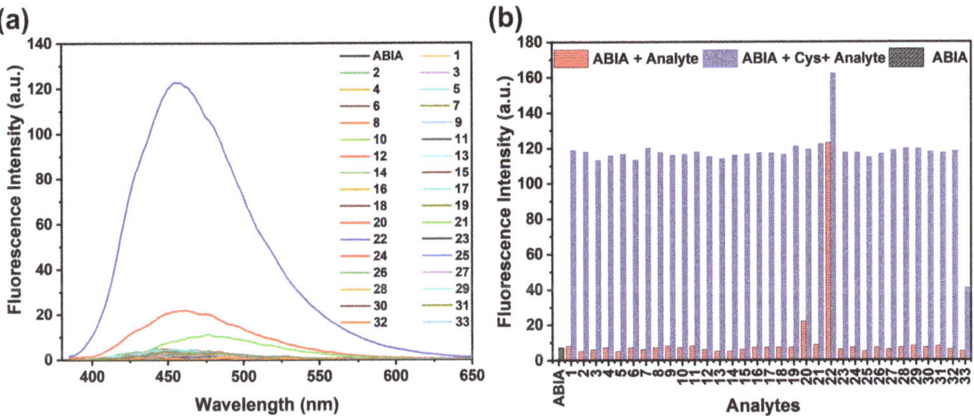

Figure 2. (**a**) Fluorescence spectra and (**b**) competitive study of **ABIA** (10 µM, λ_{ex} = 368 nm, λ_{em} = 455 nm) in the absence and presence of five equivalents of various analytes (50 µM). 1, Gly; 2, Ala; 3, Ser; 4, Pro; 5, Val; 6, Thr; 7, Ile; 8, Leu; 9, Asn; 10, Asp; 11, Gln; 12, Lys; 13, Glu; 14, Met; 15, His; 16, Phe; 17, Arg; 18, Tyr; 19, Typ; 20, Hcy; 21, Gsh; 22, Cys; 23, BSA; 24, Cl^-; 25, Br^-; 26, NO_3^-; 27, AcO^-; 28, SO_4^{2-}; 29, PO_4^{3-}; 30, Na^+; 31, Cs^+; 32, Ca^{2+}; 33, Cu^{2+}.

To estimate the minimum response time, we traced the time-dependent fluorescence intensity of **ABIA** (10 µM) for 60 min upon the addition of five equivalents of Cys, as shown in Figure S9. The fluorescence intensity almost reached the maximum (310 a.u.) in 10 min after the addition of Cys. It is important to note that the fluorescence intensity was 121 a.u. within 1 min after adding Cys, which was comparatively fast compared to the reported probes, considering the structural simplicity of **ABIA** (see the Supplemental Information Table S3). In addition, we calculated the pseudo-first-order rate constants of the reactions between **ABIA** (10 µM) and five equivalents of Cys. The calculated pseudo-first-order rate constants for Cys from the linear regression analysis of time-dependent changes in the fluorescence intensity was 5.13×10^{-3} s^{-1}, as shown in Figure S9. These results indicate that the **ABIA** shows high specificity for Cys and a short reaction time (1 min). Hence, **ABIA** is applicable for the detection and quantification of Cys in urine and live cells.

The Cys detection mechanism by **ABIA** was studied using ^{1}H NMR spectroscopy and mass spectroscopy. As shown in Figure 3, 0.5 and 1.0 equivalents of Cys were added to the 90% DMSO:0.01 M HEPES buffer solution (DMSO-d_6 and D_2O were used to make this solution) containing **ABIA**. Upon reaction, the peaks at 5.26, 5.71, and 5.97 ppm corresponding to the acrylate group of **ABIA** completely disappeared with 1.0 equivalents of Cys. These results indicate that the mercapto group in Cys undergoes the Michael addition with the double bond in the acrylate group of **ABIA.** Further, the appearance of new peaks at 2.26, 2.34, and 3.05 ppm indicates the intramolecular attack of the amine group of Cys on the carbonyl of **ABIA**, resulting in the cyclized product along with compound **3**, as shown in Figure 3b. The formation of compound **3** after the reaction of **ABIA** with Cys indicates the resultant fluorescence "turn-on" mechanism. Further proof for the proposed mechanism presented in Figure 3b was deduced from the mass spectra of a reaction product obtained by reacting equimolar **ABIA** and CYS for 60 min at 25 °C in 90% DMSO:0.01 M HEPES buffer solution. The solvent was evaporated at 40 °C under high-vacuum, and the resultant residue was prepared for the acquisition of the mass spectra by dissolving it in methanol. As presented in Figure 3b and Figure S10, the cyclization process of Cys after the reaction with **ABIA** generates a seven-membered ring (calcd. (H^+): 176.0376; found *m/z*: 176.0) and compound **3** (calcd. (H^+): 308.368; found *m/z*: 308.0) upon the de-acrylation of **ABIA**. Therefore, these results indicate the high applicability of **ABIA** as a fluorescence turn-on sensor for Cys in live cells and urine samples.

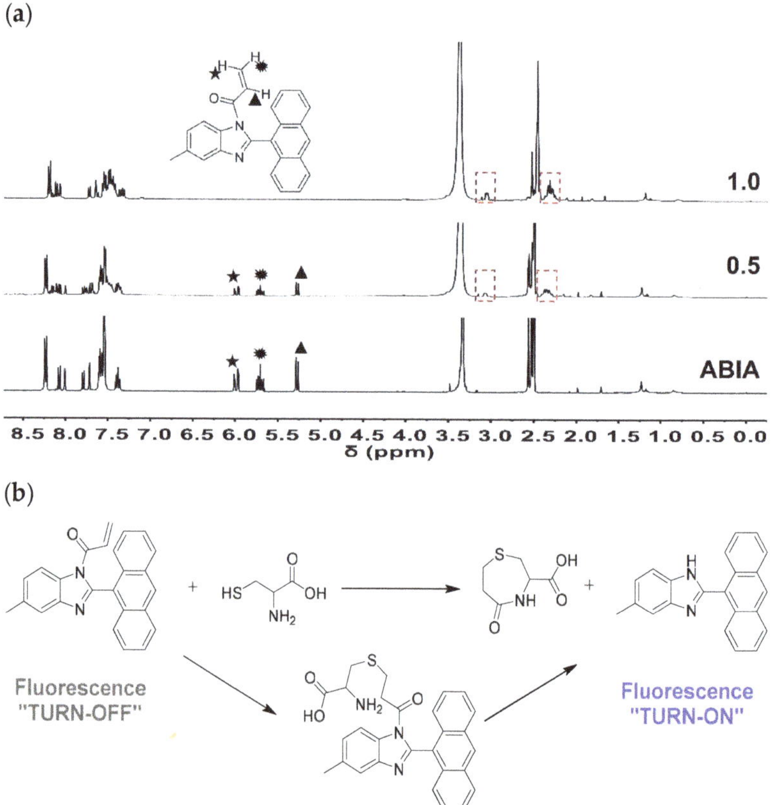

Figure 3. (**a**) Changes in ^{1}H NMR spectra of **ABIA** upon the addition of 0.5 and 1.0 equivalents of Cys in 90% DMSO-d_6:0.01 M HEPES buffer solution (D_2O); (**b**) proposed Cys detection mechanism by **ABIA**.

3.3. *The Design and Application of **CysDDev** for Detecting Cys in Simulated Human Urine*

A schematic of the **CysDDev** prototype is shown in Figure 4. The LED was arranged in a nozzle opening at a 90° angle to the sensor nozzle in the cuvette holder, allowing efficient sample illumination by UV LED and detecting emitted light by the sensor. An electric current of 15 mA was delivered to the UV LED (λ = 370 nm) under a bias of 9 V using a rechargeable Li-ion battery. The analog signals from the sensor consisted of a train of pulses correlating to the spatial position of 256 pixels corresponding to wavelengths in the range of 340–850 nm. The analog data collected from the analog pin of the sensor were digitized in a bit resolution of 10 bits using the Arduino Uno R3 microcontroller that sends the processed data to the computer. The final data of individual sample readings were saved as a Microsoft Excel file (.csv) in the pre-selected folder on the connected computer.

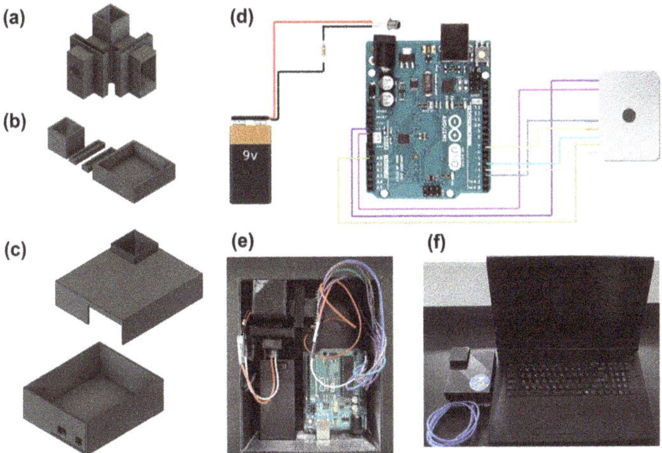

Figure 4. Schematic of various parts of the **CysDDev** (**a**): UV LED, sensor, and cuvette holder; (**b**) top lids for UV LED, sensor, cuvette holder; (**c**) the main case with top lid; (**d**) electronic components connected to the Arduino microcontroller; (**e**) assembly of **CysDDev** without the top lid; (**f**) **CysDDev** prototype connected to the laptop computer.

We compared the performance of **CysDDev** with the commercial fluorescence spectrometer using the **ABIA** (10 μM) and compound **3** (10 μM) in 90% DMSO:0.01 M HEPES buffer solution. As shown in Figure 5a, the **CysDDev** showed comparable results with the commercial fluorescence spectrometer for compound **3** and the probe **ABIA**. These results indicate that the **CysDDev** developed here can detect Cys in various solutions, including in urine.

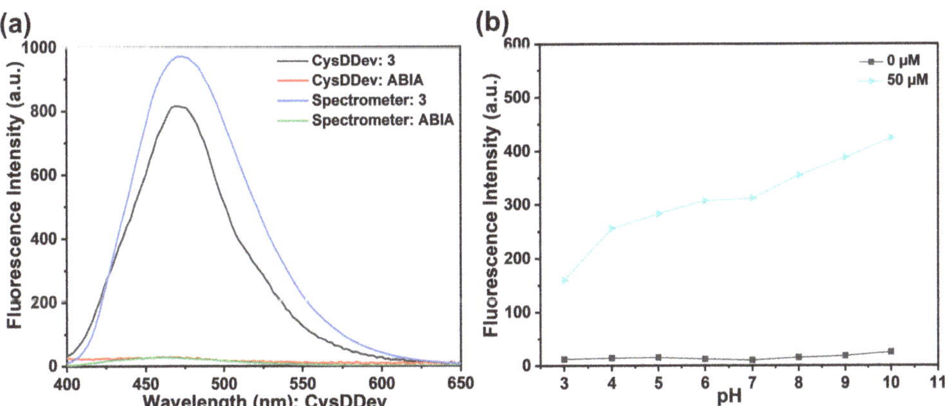

Figure 5. (**a**) Fluorescence spectra of compound **3** (10 μM, λ_{ex} = 370 nm) and **ABIA** (10 μM, λ_{ex} = 370 nm) recorded using **CysDDev** and commercial fluorescence spectrometer; (**b**) effect of pH on the detection of Cys (0, 50 μM) in simulated human urine at various pH using **ABIA** and **CysDDev** (λ_{ex} = 370 nm, λ_{em} = 455 nm).

3.4. Effect of pH on the Detection of Cys by ABIA

The effect of pH (pH = 3–10) on the detection of Cys using **ABIA** was examined by using **CysDDev** both in the absence and presence of Cys (50 μM) in simulated human urine, as shown in Figure 5b. The fluorescence intensity of the probe **ABIA** in the absence of Cys did not show a significant change in the studied pH range of 3–10, indicating that the **ABIA** has remarkable stability in acidic and alkaline environments. The fluorescence intensity increased in the presence of various concentrations of Cys from pH 3.0–10.0. The *pKa* values of the Cys for α–NH_3^+ (*pKa* = 10.25) and –SH (*pKa* = 8.0) groups dictate that below pH 8, the –SH is the only nucleophile reacting with the acrylate group of **ABIA**. However, over pH 8, the nucleophilicity of Cys α–NH_2 increases. Hence, over pH 8, the –SH and –NH_2 groups of Cys can react with **ABIA** and can enhance the de-acrylation process, resulting in higher fluorescence intensity with increasing pH. Unlike reported probes for Cys detection that require use at pH ≥ 9.0 [47–49], **ABIA** can detect Cys in the physiological pH range and the typical urine pH values of 6.0 to 7.5.

3.5. Detection of Cys Using ABIA and CysDDev in Simulated Human Urine and Real Urine Sample

Cys detection using **ABIA** and **CysDDev** was performed using urine with a pH of 7.4. As shown in Figure 6, a calibration plot (0–150 μM) was obtained by diluting a stock solution of Cys.

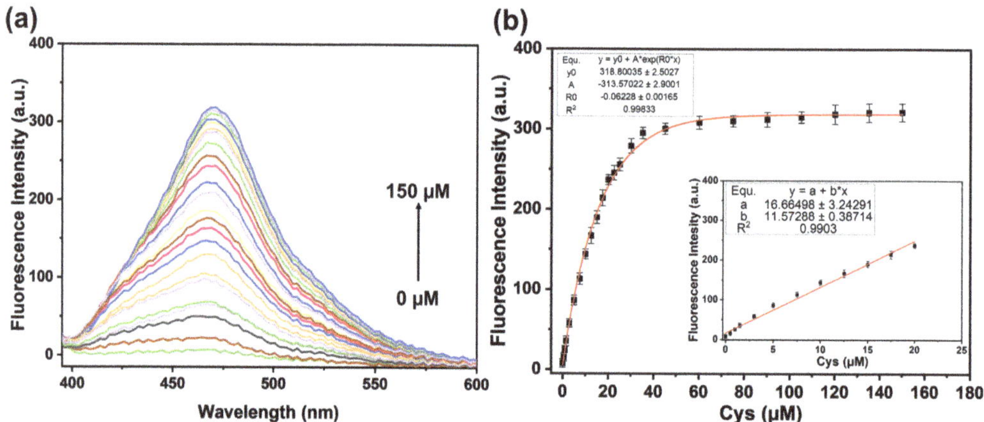

Figure 6. (**a**) Changes in fluorescence spectra of **ABIA** (10 μM, λ_{ex} = 370 nm) recorded using **CysDDev** upon the successive addition Cys (0–150 μM) in urine at pH = 7.3 (**b**) Calibration curves for the detection of Cys in urine using **ABIA** and **CysDDev** (λ_{ex} = 370 nm, λ_{em} = 455 nm) Inset is the linear regression curve in the range of 0–20 μM of Cys.

As depicted in Figure 6, the fluorescence signals of **ABIA** (10 μM) increase linearly upon the successive addition of Cys until 20 μM and then reach the plateau around 50 μM and remains the same until 150 μM. The LOD for detecting Cys by **ABIA** using **CysDDev** was found to be 16.3 nM, with a linear detection range of 0–20 μM. It is imperative to note that **ABIA** demonstrated relatively lower detection limits for Cys than the other methods presented in Table S4.

Here, we used the standard addition method to quantify the Cys in the urine sample of a healthy volunteer to minimize the interference from the urine components. The fresh urine sample without any pretreatment was spiked with 3, 6, and 9 μM of Cys, and the spiked recovery was measured using the standard curve (Figure 6b (inset)). The results of the spike test are presented in Table 1. As shown in Table 1, the probe **ABIA** and **CysDDev** can effectively detect the Cys in the real urine sample with statistically significant recovery (99.20–104.9%) and precision (RSD < 1.45%).

Table 1. Spike recovery test using real urine sample from a healthy volunteer.

Spiked Cys	Found Cys	Recovery	RSD (n = 3)
(μM)	(μM)	(%)	(%)
3	3.14	104.9	1.45
6	5.95	99.2	1.22
9	9.77	108.6	0.50

RSD, relative standard deviation.

3.6. Cell Imaging Application of **ABIA** for the Detection of Cys

The MTT assay allowed us to estimate cytotoxicity by treating cells with 0.1, 10, 25, and 50 μM of **ABIA**, Cys, and **ABIA** + Cys for 24 h using DMSO as a control. The cytotoxicity assay results are depicted in Figure 7a as the percent cell growth after treatment with **ABIA**, Cys, and **ABIA** + Cys compared to the control. There was no significant cell death even after 24 h of treatment at all tested concentrations. Therefore, 25 μM of **ABIA** was used for the cell imaging applications. As shown in Figure 7b, the control cells, cells treated Cys (500 μM), and cells pretreated with NEM (1 mM) followed by treatment with **ABIA** (25 μM) did not show significant fluorescence. However, the fluorescence intensity was observed upon the treatment of cells with Cys (500 μM) followed by **ABIA** (25 μM). Further, the fluorescence intensities were also observed in the cells pretreated with NEM (1 mM) followed sequentially by the treatment of Cys (500 μM) and **ABIA** (25 μM). These results indicate that the fluorescence turn-on probe **ABIA** has high potential in biological applications detecting intracellular Cys in the in vitro assays.

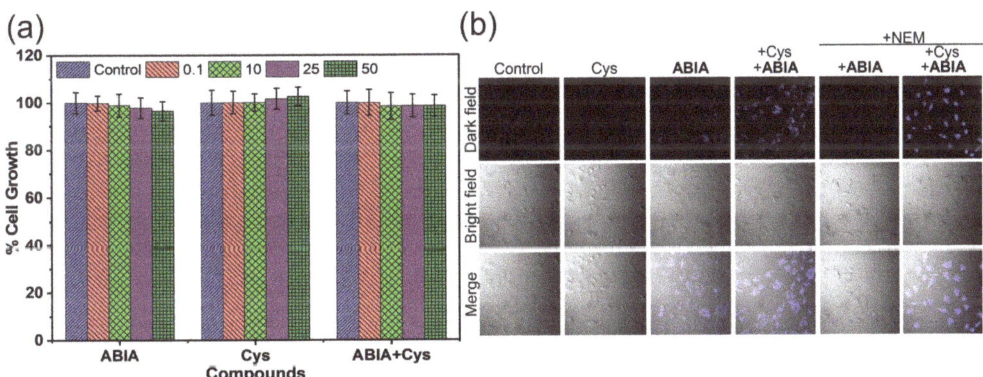

Figure 7. (**a**) Cytotoxicity's of **ABIA**, Cys, and **ABIA** + Cys on A549 cells at concentrations of 0.1, 10, 25, and 50 μM after 24 h; (**b**) Dark field, bright field, and merged images upon treatment of A549 cells with control, Cys, **ABIA**, Cys + **ABIA**, NEM + **ABIA**, and NEM + Cys + **ABIA**.

4. Discussion

There are several reports on the methods that are developed for detecting biothiols. Fluorescence turn-on sensors have attracted attention because of their high sensitivity, ease of operation, and fluorescence imaging in living cells. The Michael addition reaction [50], cyclization with aldehydes [51], conjugate addition-cyclization reaction [52], cleavage of sulfonamide, sulfonate esters [53], etc., reaction mechanisms have been reported for Cys detection among these sensors.

Most of the reported Cys detection probes are based on the coumarin scaffold with larger and more complicated structures [54,55]. Therefore, developing a synthetically accessible fluorescent probe with a simple design that shows high sensitivity and selectivity for detecting Cys can demonstrate several applications. Among the various fluorescent probes used for detecting biothiols, acrylate groups provide a fast response to the mercapto group and have a higher ability to quench fluorescence due to their strong electron-withdrawing effect [43,56,57]. Further, there are a few benzimidazole derivatives, such as fluorescent chemosensors [58,59]. However, in most cases, the benzimidazole moiety demonstrates fluorescence in the UV region (λ_{em} < 400 nm) [60], which is not suitable for live-cell imaging applications. The probe **ABIA** presented here contains an anthracene moiety, due to which it demonstrates fluorescence in the visible region, unlike other reported benzimidazole derivatives.

Compound **3**, which was presented in this manuscript, exhibited high fluorescence, upon which the subsequent reaction with acryloyl chloride resulted in the non-fluorescent **ABIA**. The directly linked acrylate group to the benzimidazole skeleton triggered an intramolecular charge transfer (ICT), resulting in a fluorescence turn-off effect. However, the Michael addition of the Cys mercapto group to the acrylate of **ABIA** restricted the ICT process and results in a fluorescence turn-on effect. **ABIA** shows selectivity for Cys over Hcy, probably because the rection of **ABIA** with Cys generates a seven-membered ring. In comparison, a reaction of **ABIA** with Hcy would produce an eight-membered ring. The strain in an eight-membered ring is much higher than that in a seven-membered ring; thus, a favorable seven-membered ring product forms easily. **ABIA** demonstrated high selectivity and comparatively fast response time for Cys detection (see the Supporting Information Table S3). The live-cell imaging study presented here provides evidence that **ABIA** has high applicability for in vitro bioimaging for Cys.

Reported probes were used for either the bioimaging or detection of Cys in serum samples. Only a few reports elaborate upon Cys detection in urine. However, the probe **ABIA** presented here was evaluated for its dual application for detecting Cys in live cells and urine. The **CysDDev** explained here allows the efficient detection of Cys in simulated human urine. The LOD for Cys detection by the probe **ABIA** using **CysDDev** was found to be of 16.3 nM. The **CysDDev** and **ABIA** demonstrated excellent clinical applicability in terms of LOD for Cys detection compared to reported methods (see the Supporting Information Table S4). Though the further evaluation of **CysDDev** for clinical application is required, these initial results indicate that **CysDDev** is an excellent portable tool that is deployable in resource-limited settings. The portability and the ease of use make **CysDDev** a perfect point-of-care device for evaluating clinical samples.

5. Conclusions

The novel fluorescence "off-on" probe **ABIA** to detect the mercapto-group containing amino acids was designed and synthesized. In addition, a portable device, **CysDDev**, was also designed and developed to measure the Cys in urine. The probe **ABIA** and **CysDDev** exhibited highly sensitive, simple, rapid, and inexpensive cysteine detection ability. The acrylate group in **ABIA** reacted with the mercapto group in Cys by means of the Michael addition reaction mechanism. The Michael addition reaction was followed by selective intramolecular nucleophilic substitution, ensuring selective detection of Cys over Hcy and GSH. The response time of 1 min and a detection limit of 16.3 nM signifies the high clinical applicability of **ABIA** and **CysDDev**. Further, **ABIA** also demonstrated its utility in detecting intracellular cysteine, making it an excellent probe for bio-imaging assay.

These results suggest that the **ABIA** and **CysDDev** will provide new insights into developing convenient approaches in Cys and other biomarker detections through simple, accurate, and portable health products for potential biological applications.

Supplementary Materials: The following are available online at https://www.mdpi.com/article/10.3390/bios11110420/s1, Figure S1: ^{1}H-NMR spectrum of compound 3, Figure S2: ^{13}C-NMR spectrum of compound 3, Figure S3: HR-mass spectrum of compound 3, Figure S4: ^{1}H-NMR spectrum of ABIA, Figure S5: ^{13}C-NMR spectrum of ABIA, Figure S6: HR-mass spectrum of ABIA, Figure S7: Structures of the probe ABIA tautomer's A and B, Figure S8: (a) UV-vis absorbance spectra, (b) Fluorescence spectra of compound 3 in 10, 50, 70, and 90% DMSO:0.01 M HEPES buffer solution (λ_{ex} = 368 nm), Figure S9: (a), (b)Time-dependent fluorescence response and (c) pseudo-first-order kinetic plots of ABIA (10 µM) reaction with Cys (50 µM) in 90% DMSO:0.01 M HEPES buffer solution (λ_{ex} = 368 nm,

λ_{em} = 455 nm), Figure S10: Mass spectra of a reaction product obtained by reacting equimolar ABIA and Cys for 60 min at 25 °C, Table S1: ^{1}H NMR and ^{13}C NMR chemical shifts (δ ppm) of the ABIA tautomer's A and B, Table S1. ^{1}H NMR and ^{13}C NMR chemical shifts (δ ppm) of the ABIA tautomer's A and B; Table S2: Photophysical properties of compound 3 (10 μM) in varying percentages of DMSO:0.01 M HEPES buffer solution, Table S3: A comparison of response time for cysteine detection by various methods, Table S4: A comparison of different probes used for cysteine detection.

Author Contributions: Conceptualization, S.B.N.; methodology, G.S.Y., I.-h.S., T.K., P.B.S., S.D.W., S.-m.P. and S.B.N.; software, T.K.; formal analysis, G.S.Y., I.-h.S., S.D.W., T.K. and S.B.N.; investigation, G.S.Y., I.-h.S., T.K., S.D.W. and S.-m.P.; data curation, G.S.Y., I.-h.S. and S.B.N.; writing—original draft preparation, G.S.Y. and S.B.N.; writing—review and editing, P.B.S., T.K. and S.B.N.; supervision, S.B.N.; project administration, P.B.S. and S.B.N.; funding acquisition, S.B.N. All authors have read and agreed to the published version of the manuscript.

Funding: Hallym University Research Fund (HRF-202012-009) supported this research.

Institutional Review Board Statement: Not applicable.

Informed Consent Statement: Not applicable.

Data Availability Statement: The data presented in this study are available upon request from the corresponding author.

Conflicts of Interest: The authors declare no conflict of interest.

References

1. Serpa, J. Cysteine as a Carbon Source, a Hot Spot in Cancer Cells Survival. *Front. Oncol.* **2020**, *10*, 947. [CrossRef] [PubMed]
2. Yu, L.; Teoh, S.T.; Ensink, E.; Ogrodzinski, M.P.; Yang, C.; Vazquez, A.I.; Lunt, S.Y. Cysteine catabolism and the serine biosynthesis pathway support pyruvate production during pyruvate kinase knockdown in pancreatic cancer cells. *Cancer Metab.* **2019**, *7*, 13. [CrossRef] [PubMed]
3. Wu, G.; Fang, Y.-Z.; Yang, S.; Lupton, J.R.; Turner, N.D. Glutathione Metabolism and Its Implications for Health. *J. Nutr.* **2004**, *134*, 489–492. [CrossRef] [PubMed]
4. Liu, X.; Zhang, S.; Whitworth, R.J.; Stuart, J.J.; Chen, M.-S. Unbalanced Activation of Glutathione Metabolic Pathways Suggests Potential Involvement in Plant Defense against the Gall Midge Mayetiola destructor in Wheat. *Sci. Rep.* **2015**, *5*, 8092. [CrossRef]
5. Blom, H.J.; Smulders, Y. Overview of homocysteine and folate metabolism. With special references to cardiovascular disease and neural tube defects. *J. Inherit. Metab. Dis.* **2011**, *34*, 75–81. [CrossRef]
6. Hasan, T.; Arora, R.; Bansal, A.K.; Bhattacharya, R.; Sharma, G.S.; Singh, L.R. Disturbed homocysteine metabolism is associated with cancer. *Exp. Mol. Med.* **2019**, *51*, 1–13. [CrossRef]
7. Daher, B.; Vučetić, M.; Pouysségur, J. Cysteine Depletion, a Key Action to Challenge Cancer Cells to Ferroptotic Cell Death. *Front. Oncol.* **2020**, *10*, 723. [CrossRef]
8. Dröge, W. Oxidative stress and ageing: Is ageing a cysteine deficiency syndrome? *Philos. Trans. R. Soc. B Biol. Sci.* **2005**, *360*, 2355–2372. [CrossRef]
9. Rehman, T.; Shabbir, M.A.; Inam-Ur-Raheem, M.; Manzoor, M.F.; Ahmad, N.; Liu, Z.-W.; Ahmad, M.H.; Siddeeg, A.; Abid, M.; Aadil, R.M. Cysteine and homocysteine as biomarker of various diseases. *Food Sci. Nutr.* **2020**, *8*, 4696–4707. [CrossRef]
10. Rhodes, H.L.; Yarram-Smith, L.; Rice, S.J.; Tabaksert, A.; Edwards, N.; Hartley, A.; Woodward, M.N.; Smithson, S.L.; Tomson, C.; Welsh, G.I.; et al. Clinical and Genetic Analysis of Patients with Cystinuria in the United Kingdom. *Clin. J. Am. Soc. Nephrol.* **2015**, *10*, 1235. [CrossRef]
11. Otter, D.E. Standardised methods for amino acid analysis of food. *Br. J. Nutr.* **2012**, *108*, S230–S237. [CrossRef]
12. Maheshwari, H.; Vilà, N.; Herzog, G.; Walcarius, A. Selective Detection of Cysteine at a Mesoporous Silica Film Electrode Functionalized with Ferrocene in the Presence of Glutathione. *ChemElectroChem* **2020**, *7*, 2095–2101. [CrossRef]
13. Jongjinakool, S.; Palasak, K.; Bousod, N.; Teepoo, S. Gold Nanoparticles-based Colorimetric Sensor for Cysteine Detection. *Energy Procedia* **2014**, *56*, 10–18. [CrossRef]
14. Tajik, S.; Dourandish, Z.; Jahani, P.M.; Sheikhshoaie, I.; Beitollahi, H.; Shahedi Asl, M.; Jang, H.W.; Shokouhimehr, M. Recent developments in voltammetric and amperometric sensors for cysteine detection. *RSC Adv.* **2021**, *11*, 5411–5425. [CrossRef]
15. Han, C.; Yang, H.; Chen, M.; Su, Q.; Feng, W.; Li, F. Mitochondria-Targeted Near-Infrared Fluorescent Off–On Probe for Selective Detection of Cysteine in Living Cells and in Vivo. *ACS Appl. Mater. Interfaces* **2015**, *7*, 27968–27975. [CrossRef]
16. Yang, Z.; Cao, J.; He, Y.; Yang, J.H.; Kim, T.; Peng, X.; Kim, J.S. Macro-/micro-environment-sensitive chemosensing and biological imaging. *Chem. Soc. Rev.* **2014**, *43*, 4563–4601. [CrossRef]
17. Chan, J.; Dodani, S.C.; Chang, C.J. Reaction-based small-molecule fluorescent probes for chemoselective bioimaging. *Nat. Chem.* **2012**, *4*, 973–984. [CrossRef]
18. Zhou, M.; Ding, J.; Guo, L.-P.; Shang, Q.-K. Electrochemical Behavior of l-Cysteine and Its Detection at Ordered Mesoporous Carbon-Modified Glassy Carbon Electrode. *Anal. Chem.* **2007**, *79*, 5328–5335. [CrossRef]

19. Zhang, W.; Zhao, X.; Gu, W.; Cheng, T.; Wang, B.; Jiang, Y.; Shen, J. A novel naphthalene-based fluorescent probe for highly selective detection of cysteine with a large Stokes shift and its application in bioimaging. *New J. Chem.* **2018**, *42*, 18109–18116. [CrossRef]
20. Fu, Z.-H.; Han, X.; Shao, Y.; Fang, J.; Zhang, Z.-H.; Wang, Y.-W.; Peng, Y. Fluorescein-Based Chromogenic and Ratiometric Fluorescence Probe for Highly Selective Detection of Cysteine and Its Application in Bioimaging. *Anal. Chem.* **2017**, *89*, 1937–1944. [CrossRef]
21. Dong, B.; Lu, Y.; Zhang, N.; Song, W.; Lin, W. Ratiometric Imaging of Cysteine Level Changes in Endoplasmic Reticulum during H2O2-Induced Redox Imbalance. *Anal. Chem.* **2019**, *91*, 5513–5516. [CrossRef]
22. Yu, S.; Nimse, S.B.; Kim, J.; Song, K.-S.; Kim, T. Development of a Lateral Flow Strip Membrane Assay for Rapid and Sensitive Detection of the SARS-CoV-2. *Anal. Chem.* **2020**, *92*, 14139–14144. [CrossRef]
23. Tang, B.; Xing, Y.; Li, P.; Zhang, N.; Yu, F.; Yang, G. A Rhodamine-Based Fluorescent Probe Containing a Se−N Bond for Detecting Thiols and Its Application in Living Cells. *J. Am. Chem. Soc.* **2007**, *129*, 11666–11667. [CrossRef]
24. Yuan, L.; Lin, W.; Zhao, S.; Gao, W.; Chen, B.; He, L.; Zhu, S. A Unique Approach to Development of Near-Infrared Fluorescent Sensors for in Vivo Imaging. *J. Am. Chem. Soc.* **2012**, *134*, 13510–13523. [CrossRef]
25. Yang, X.; Guo, Y.; Strongin, R.M. Conjugate Addition/Cyclization Sequence Enables Selective and Simultaneous Fluorescence Detection of Cysteine and Homocysteine. *Angew. Chem. Int. Ed.* **2011**, *50*, 10690–10693. [CrossRef]
26. Long, Y.-T.; Kong, C.; Li, D.-W.; Li, Y.; Chowdhury, S.; Tian, H. Ultrasensitive Determination of Cysteine Based on the Photocurrent of Nafion-Functionalized CdS–MV Quantum Dots on an ITO Electrode. *Small* **2011**, *7*, 1624–1628. [CrossRef]
27. Wang, S.; Zhang, Q.; Chen, S.; Wang, K.-P.; Hu, Z.-Q. A diazabenzoperylene derivative as ratiometric fluorescent probe for cysteine with super large Stokes shift. *Anal. Bioanal. Chem.* **2020**, *412*, 2687–2696. [CrossRef]
28. Lim, S.; Escobedo, J.O.; Lowry, M.; Xu, X.; Strongin, R. Selective fluorescence detection of cysteine and N-terminal cysteine peptide residues. *Chem. Commun.* **2010**, *46*, 5707–5709. [CrossRef]
29. He, L.; Yang, X.; Xu, K.; Lin, W. Improved Aromatic Substitution–Rearrangement-Based Ratiometric Fluorescent Cysteine-Specific Probe and Its Application of Real-Time Imaging under Oxidative Stress in Living Zebrafish. *Anal. Chem.* **2017**, *89*, 9567–9573. [CrossRef]
30. Liu, B.; Wang, J.; Zhang, G.; Bai, R.; Pang, Y. Flavone-Based ESIPT Ratiometric Chemodosimeter for Detection of Cysteine in Living Cells. *ACS Appl. Mater. Interfaces* **2014**, *6*, 4402–4407. [CrossRef]
31. Zhou, J.; Yu, C.; Li, Z.; Peng, P.; Zhang, D.; Han, X.; Tang, H.; Wu, Q.; Li, L.; Huang, W. A rapid and highly selective paper-based device for high-throughput detection of cysteine with red fluorescence emission and a large Stokes shift. *Anal. Methods* **2019**, *11*, 1312–1316. [CrossRef]
32. Di Nonno, S.; Ulber, R. Smartphone-based optical analysis systems. *Analyst* **2021**, *146*, 2749–2768. [CrossRef] [PubMed]
33. Shariati, S.; Khayatian, G. The colorimetric and microfluidic paper-based detection of cysteine and homocysteine using 1,5-diphenylcarbazide-capped silver nanoparticles. *RSC Adv.* **2021**, *11*, 3295–3303. [CrossRef]
34. Mehta, S.M.; Mehta, S.; Muthurajan, H.; D'Souza, J.S. Vertical flow paper-based plasmonic device for cysteine detection. *Biomed. Microdevices* **2019**, *21*, 55. [CrossRef]
35. Ma, W.-W.; Wang, M.-Y.; Yin, D.; Zhang, X. Facile preparation of naphthol AS-based fluorescent probe for highly selective detection of cysteine in aqueous solution and its imaging application in living cells. *Sens. Actuators B Chem.* **2017**, *248*, 332–337. [CrossRef]
36. Long, G.L.; Winefordner, J.D. Limit of Detection A Closer Look at the IUPAC Definition. *Anal. Chem.* **1983**, *55*, 712A–724A. [CrossRef]
37. Lee, J.-S.; Warkad, S.D.; Shinde, P.B.; Kuwar, A.; Nimse, S.B. A highly selective fluorescent probe for nanomolar detection of ferric ions in the living cells and aqueous media. *Arab. J. Chem.* **2020**, *13*, 8697–8707. [CrossRef]
38. Song, I.-H.; Torawane, P.; Lee, J.-S.; Warkad, S.D.; Borase, A.; Sahoo, S.K.; Nimse, S.B.; Kuwar, A. The detection of Al^{3+} and Cu^{2+} ions using isonicotinohydrazide-based chemosensors and their application to live-cell imaging. *Mater. Adv.* **2021**, *2*, 6306–6314. [CrossRef]
39. Yaragorla, S.; Vijaya Babu, P. Oxidative Csp3-H functionalization of 2-methylazaarenes: A practical synthesis of 2-azaarenyl-benzimidazoles and benzothiazoles. *Tetrahedron Lett.* **2017**, *58*, 1879–1882. [CrossRef]
40. Baig, M.F.; Shaik, S.P.; Nayak, V.L.; Alarifi, A.; Kamal, A. Iodine-catalyzed Csp3-H functionalization of methylhetarenes: One-pot synthesis and cytotoxic evaluation of heteroarenyl-benzimidazoles and benzothiazole. *Bioorganic Med. Chem. Lett.* **2017**, *27*, 4039–4043. [CrossRef]
41. Kadri, H.; Matthews, C.S.; Bradshaw, T.D.; Stevens, M.F.; Westwell, A.D. Synthesis and antitumour evaluation of novel 2-phenylbenzimidazoles. *J. Enzyme Inhib. Med. Chem.* **2008**, *23*, 641–647. [CrossRef]
42. Lee, I.-S.H.; Jeoung, E.H.; Lee, C.K. Synthesis and tautomerism of 2-aryl- and 2-heteroaryl derivatives of benzimidazole. *J. Heterocycl. Chem.* **1996**, *33*, 1711–1716. [CrossRef]
43. Dai, Y.; Zheng, Y.; Xue, T.; He, F.; Ji, H.; Qi, Z. A novel fluorescent probe for rapidly detection cysteine in cystinuria urine, living cancer/normal cells and BALB/c nude mice. *Spectrochim. Acta Part A Mol. Biomol. Spectrosc.* **2020**, *225*, 117490. [CrossRef]
44. Toro-Roman, V.; Siquier-Coll, J.; Bartolome, I.; Grijota, F.J.; Munoz, D.; Maynar-Marino, M. Copper concentration in erythrocytes, platelets, plasma, serum and urine: Influence of physical training. *J. Int. Soc. Sports Nutr.* **2021**, *18*, 28. [CrossRef]
45. Kaniowska, E.; Chwatko, G.; Glowacki, R.; Kubalczyk, P.; Bald, E. Urinary excretion measurement of cysteine and homocysteine in the form of their S-pyridinium derivatives by high-performance liquid chromatography with ultraviolet detection. *J. Chromatogr. A* **1998**, *798*, 27–35. [CrossRef]
46. Zelikovic, I. Chapter 19—Hereditary Tubulopathies. In *Nephrology and Fluid/electrolyte Physiology*, 3rd ed.; Oh, W., Baum, M., Eds.; Elsevier: Philadelphia, PA, USA, 2019; pp. 315–344.
47. Rusin, O.; St Luce, N.N.; Agbaria, R.A.; Escobedo, J.O.; Jiang, S.; Warner, I.M.; Dawan, F.B.; Lian, K.; Strongin, R.M. Visual Detection of Cysteine and Homocysteine. *J. Am. Chem. Soc.* **2004**, *126*, 438–439. [CrossRef]

48. Xu, Y.; Li, B.; Han, P.; Sun, S.; Pang, Y. Near-infrared fluorescent detection of glutathione via reaction-promoted assembly of squaraine-analyte adducts. *Analyst* **2013**, *138*, 1004–1007. [CrossRef]
49. Wei, M.; Yin, P.; Shen, Y.; Zhang, L.; Deng, J.; Xue, S.; Li, H.; Guo, B.; Zhang, Y.; Yao, S. A new turn-on fluorescent probe for selective detection of glutathione and cysteine in living cells. *Chem. Commun.* **2013**, *49*, 4640–4642. [CrossRef]
50. Chen, Z.; Sun, Q.; Yao, Y.; Fan, X.; Zhang, W.; Qian, J. Highly sensitive detection of cysteine over glutathione and homo-cysteine: New insight into the Michael addition of mercapto group to maleimide. *Biosens. Bioelectron.* **2017**, *91*, 553–559. [CrossRef]
51. Wang, P.; Liu, J.; Lv, X.; Liu, Y.; Zhao, Y.; Guo, W. A Naphthalimide-Based Glyoxal Hydrazone for Selective Fluorescence Turn-On Sensing of Cys and Hcy. *Org. Lett.* **2012**, *14*, 520–523. [CrossRef]
52. Zhang, J.; Wang, J.; Liu, J.; Ning, L.; Zhu, X.; Yu, B.; Liu, X.; Yao, X.; Zhang, H. Near-Infrared and Naked-Eye Fluorescence Probe for Direct and Highly Selective Detection of Cysteine and Its Application in Living Cells. *Anal. Chem.* **2015**, *87*, 4856–4863. [CrossRef]
53. Sun, Q.; Yang, S.-H.; Wu, L.; Yang, W.-C.; Yang, G.-F. A Highly Sensitive and Selective Fluorescent Probe for Thiophenol Designed via a Twist-Blockage Strategy. *Anal. Chem.* **2016**, *88*, 2266–2272. [CrossRef]
54. Bugaenko, D.I.; Karchava, A.V.; Yunusova, Z.A.; Yurovskaya, M.A. Fluorescent probes on the basis of coumarin derivatives for determining biogenic thiols and thiophenols. *Chem. Heterocycl. Compd.* **2019**, *55*, 483–489. [CrossRef]
55. Cao, D.; Liu, Z.; Verwilst, P.; Koo, S.; Jangjili, P.; Kim, J.S.; Lin, W. Coumarin-Based Small-Molecule Fluorescent Chemosensors. *Chem. Rev.* **2019**, *119*, 10403–10519. [CrossRef]
56. Liu, G.; Liu, D.; Han, X.; Sheng, X.; Xu, Z.; Liu, S.H.; Zeng, L.; Yin, J. A hemicyanine-based colorimetric and ratiometric fluorescent probe for selective detection of cysteine and bioimaging in living cell. *Talanta* **2017**, *170*, 406–412. [CrossRef]
57. Qi, Y.; Huang, Y.; Li, B.; Zeng, F.; Wu, S. Real-Time Monitoring of Endogenous Cysteine Levels In Vivo by near-Infrared Turn-on Fluorescent Probe with Large Stokes Shift. *Anal. Chem.* **2018**, *90*, 1014–1020. [CrossRef]
58. Tang, L.; Shi, J.; Huang, Z.; Yan, X.; Zhang, Q.; Zhong, K.; Hou, S.; Bian, Y. An ESIPT-based fluorescent probe for selective detection of homocysteine and its application in live-cell imaging. *Tetrahedron Lett.* **2016**, *57*, 5227–5231. [CrossRef]
59. Odyniec, M.L.; Park, S.-J.; Gardiner, J.E.; Webb, E.C.; Sedgwick, A.C.; Yoon, J.; Bull, S.D.; Kim, H.M.; James, T.D. A fluorescent ESIPT-based benzimidazole platform for the ratiometric two-photon imaging of ONOO−In vitro and ex vivo. *Chem. Sci.* **2020**, *11*, 7329–7334. [CrossRef]
60. Sharma, S.; Pradeep, C.P.; Dhir, A. Benzimidazole Based 'Turn on' Fluorescent Chemodosimeter for Zinc Ions in Mixed Aqueous Medium. *J. Fluoresc.* **2016**, *26*, 1439–1445. [CrossRef]

 *biosensors* MDPI

Review

Recent Progress in Electrochemical Immunosensors

JeeYoung Kim [1,2,3] and Min Park [1,2,3,*]

[1] Major in Materials Science and Engineering, Hallym University, Chuncheon 24252, Gangwon-do, Korea; jyoung@hallym.ac.kr
[2] Cooperative Course of Nano-Medical Device Engineering, Hallym University, Chuncheon 24252, Gangwon-do, Korea
[3] Integrative Materials Research Institute, Hallym University, Chuncheon 24252, Gangwon-do, Korea
[*] Correspondence: minpark@hallym.ac.kr

Abstract: Biosensors used for medical diagnosis work by analyzing physiological fluids. Antibodies have been frequently used as molecular recognition molecules for the specific binding of target analytes from complex biological solutions. Electrochemistry has been introduced for the measurement of quantitative signals from transducer-bound analytes for many reasons, including good sensitivity. Recently, numerous electrochemical immunosensors have been developed and various strategies have been proposed to detect biomarkers. In this paper, the recent progress in electrochemical immunosensors is reviewed. In particular, we focused on the immobilization methods using antibodies for voltammetric, amperometric, impedimetric, and electrochemiluminescent immunosensors.

Keywords: immunosensors; electrochemical immunosensors; biosensors; voltammetric immunosensors; amperometric immunosensors; impedimetric immunosensors; eletrochemiluminescent immunosensors

Citation: Kim, J.; Park, M. Recent Progress in Electrochemical Immunosensors. *Biosensors* **2021**, *11*, 360. https://doi.org/10.3390/bios11100360

Received: 31 August 2021
Accepted: 27 September 2021
Published: 29 September 2021

1. Introduction

A biosensor is an analytical device that can specifically quantify the target analyte in a physiological sample, such as blood, serum, plasma, cerebrospinal fluid, urine, and interstitial fluid [1,2]. Biosensors have the advantages of portability (because they are miniature), simplicity, automation, cost-effectiveness, high stability, and a short detection time [3–7]. Moreover, biosensors can provide real-time responses, and consequently, they are suitable for point-of-care testing [8]. Biosensors are used in various fields, including clinical diagnosis, agriculture, the food industry, environmental monitoring, and quality control [9].

Biosensors are normally composed of three main parts: a molecular recognition layer, a transducer, and a signal generator [2,10]. The molecular recognition layer of biosensors is distinct from that of other sensors because the sample analysis generally involves a complicated mixture in biosensors. The molecular recognition layer is produced by the immobilization of bioreceptors that have specific binding properties to target analytes [11,12]. Antibodies are one of the most widely used molecules because of their high specificity, affinity, and ease of production; therefore, they have been utilized in various applications such as chromatography, therapy, diagnosis, immunoassays, and biosensors [13,14]. Antigen–antibody binding-based biosensors are called immunosensors, and the antibody layers are called immunoaffinity layers.

There are various types of immunosensors depending on the transducer type, such as thermal, optical, magnetic, piezoelectric, fluorescent, and electrochemical biosensors [15,16]. Electrochemical biosensors measure electrochemical signals using a chemically modified electrode [17] and they have the distinct advantage of simplicity [18]. As described previously, electrochemical biosensors can be produced by the fabrication of chemically modified electrodes. Owing to their simplicity, electrochemical biosensors have the characteristics of good durability, easy miniaturization, small analyte volumes, and the ability to be integrated with fluidic systems. Another key advantage of electrochemical biosensors is their broad detection range and excellent detection limits [18]. Therefore, electrochemical biosensors, including electrochemical immunosensors, have been actively studied.

In this review, various types of electrochemical immunosensors are introduced and discussed. In particular, the current information on voltammetric, amperometric, impedimetric, and ECL immunosensors is summarized (abbreviations are described in Table S1). This review focused on electrochemical immunosensors, and recent studies are discussed and summarized. In particular, the formation of an immunoaffinity layers for the fabrication of the electrochemical immunosensors is discussed. Recently, a couple of reviews focused on electrochemical biosensors have been published. Li et al. especially focused on CMOS and summarized the instrumentation of CMOS biosensors [19]. Juska and Pemble summarized the evolution of the electrochemical biosensor [20]. While this review focused on enzymatic electrochemical biosensors, especially glucose biosensors, our review focuses on antibody-based biosensors. Zhang and Yang focused on materials and techniques for the fabrication of electrochemical biosensors [21]. They summarized materials used for the electrode, immobilizing molecules, covering membrane and immobilization of molecular recognition molecules, and discussed various construction techniques for the preparation of biosensors. Schmidt-Speicher and Länge summarized electrochemical biosensors with integrated microfluidics [22]. The difference from the reviews discussed above is that our review focuses on immunosensors employing antibody immobilization techniques in the transducer.

2. Voltammetric Immunosensors

In voltammetric immunosensors, the current is obtained from the analyte measurements when the potential is changed. The potential between the working and reference electrodes changes over time, and the current generated between the working and counter electrodes vi the electrochemical reaction of the analytes is measured [23]. The signal originates from the oxidoreduction processes at the surface of the working electrode, caused by the electroactive species. According to the function of the applied potential, voltammetry is classified as LSV, DPV, SWV, or CV [24–27].

In voltammetric biosensors, the signal is normally collected from the peak or plateau, and the amount of the target analyte can be quantified as a proportion of the height of the peak. Voltammetry is an electrochemical analytical method for immunosensors that is widely used owing to its minimal noise, good sensitivity, and applicability [28,29]. In addition, voltammetry can be utilized for multiplex detection based on the different positions of the oxidoreduction peaks. The immune reaction cannot generate a voltammetric signal, so signal-generating molecules are essential for the fabrication of voltammetric immunosensors. From this reason, one of the major challenges in the development of voltammetric biosensors is the synthesis of highly sensitive signal-generating or -amplifying composites for the immunoaffinity layer or detection antibodies. In this section, voltammetric immunosensors employing various antibody immobilization methods are discussed, and examples are presented.

2.1. Physical Adsorption of Antibodies

The formation of an immunoaffinity layer during the fabrication of electrochemical immunosensors, especially the immobilization of antibodies on the electrode, is an important process. Carbon-based materials are among the most widely used materials for electrode modification. Ribeiro et al. developed a reusable immunosensor by immobilizing antibodies on carbon-modified electrodes through non-covalent interactions and utilized them to detect CRP [30]. As shown in Figure 1A, the graphite electrode was first modified with graphene by electrochemical reduction using GO. Then, the electrode was coated with electropolymerized polytyramine. Antibodies against CRP were immobilized on the electrode through the charge interaction between the ammonium ion of polytyramine and the carboxyl group of the antibodies. After treatment with the analyte, CRP, the signal was measured using DPV. The fabricated immunosensor showed an LOQ of 1.25 μg/L, with a linear range of 1.09–100 μg/L, and it could be reused four times. Trindade et al. fabricated a label-free immunosensor for CysC detection [31]. The gold electrode was modified using an AF-functionalized GO nanocomposite, and then it was coated with PEI. Subsequently, antibodies against CysC were immobilized by simple drop casting. After the immunoaffinity layer was formed, CysC was quantified using SWV. From the measurement, the LOD and linear range were calculated to be 0.03 ng/mL and 0.1–1000 ng/mL, respectively.

Gold is another widely used material for electrode modification and antibody immobilization because of its biocompatibility [32]. Zhou et al. developed an electrochemical immunosensor based on antibody-immobilized AuNPs [33]. The transducer used in that study was composed of antibodies immobilized on the AuNP layer for the detection of PSA. HP5-decorated AuNPs were used for signal amplification. As shown in Figure 1B, the hybrid HP5–AuNP was modified by coating it with graphitic carbon nitride through $\pi-\pi$ interactions between HP5 and graphitic carbon nitride. After the addition of methylene blue, the detection antibodies were conjugated with the nanocomposite through physical adsorption. The detection antibody conjugate was added after antigen treatment to form a sandwich-type immunocomplex, and the amplified signal was measured using DPV. The fabricated biosensor showed a 0.12 pg/mL LOD with a linear range of 0.0005–10.00 ng/mL. Suresh et al. utilized enzyme-conjugated detection antibodies for signal amplification to detect PSA [34]. AuNP was electrochemically synthesized on a CS-coated electrode, and capture antibodies were immobilized on the electrode through physical adsorption. After analyte treatment, HRP-labeled detection antibodies were added, and the signal was amplified by the reduction of hydrogen peroxide by HRP. The signal was measured using SWV, and the LOD and linear range of the developed immunosensor were calculated to be 0.001 ng/mL and 1–18 ng/mL, respectively. Zhang et al. similarly developed a PSA immunosensor based on a sandwich immunocomplex [35] and they prepared an immunoaffinity layer by sequentially coating the electrode with PANI, AuNPs, and antibodies. Then, methylene blue was encapsulated in mesoporous silica NPs, which were coated with PDA, and then detection antibodies were conjugated to the prepared particles. After the sandwich immunocomplex was formed, acid was added to release the encapsulated methylene blue, and the released methylene blue increased the signal of SWV. The fabricated immunosensor showed an LOD of 1.25 fg/mL with a linear range of 0.01–100 ng/mL. Li et al. fabricated a dual-mode immunosensor to detect procalcitonin, where antibodies were immobilized on a simple electrolytic gold matrix [36]. For signal generation, $CuCo_2O_4$ hollow spheres were coated with AuNPs, and detection antibodies were immobilized on the spheres. After the sandwich immunocomplex was formed, the signal was measured using dual mode, SWV and chronoamperometry, and the LODs of the developed immunosensor were calculated to be 82.6 fg/mL for SWV and 95.4 fg/mL for chronoamperometry, with 0.0001–50 ng/mL as the linear range.

Various studies have utilized carbon-based materials together with gold. Amani et al. developed an immunosensor utilizing graphene and AuNPs for the detection of NSE [37]. The electrode was sequentially coated with graphene, PPD, and AuNPs. Subsequently, the antibodies were immobilized on the AuNPs through physical adsorption. After NSE treatment, the signal was measured using DPV, and the fabricated immunosensor showed an LOD of 0.3 ng/mL and a linear range of 1.0–1000 ng/mL. Khoshroo et al. fabricated a CA15-3 biosensor based on a cobalt sulfide-graphene nanocomposite [38]. In that study, AuNPs were coated on a CoS_2–graphene modified electrode, and antibodies were immobilized on the AuNPs. CA15-3 was quantified using DPV, and the LOD and linear range were calculated to be 0.03 U/mL and 0.1 150.0 U/mL, respectively. Zhao et al. introduced platinum into an immunosensor to detect NMP22 [39]. They fabricated MOFs using AuNPs and PtNPs, and these MOFs were coated on a graphene-modified electrode. After NMP22 treatment, the signal was measured using DPV, and the fabricated immunosensor showed an LOD of 1.7 pg/mL with a linear range of 0.005–20 ng/mL. Assari et al. used CNTs to develop immunosensors for the detection of PSA [40]. They sequentially coated electrodes with MWCNTs, PANI, and AuNPs. Then, capture antibodies were immobilized on the modified electrode, and the signal measured with the immunosensor using DPV showed a calculated LOD and linear range of 0.5 pg/mL and 1.66 ag/mL–1.3 ng/mL, respectively. Chen et al. fabricated biosensors utilizing MWCNTs and AuNPs to detect PTH [41]. After sequentially coating the MWCNTs and AuNPs, antibodies against PTH were immobilized on AuNPs, and PTH was quantified using DPV and SWV (Figure 1C). The fabricated immunosensor showed an LOD of 886 fg/mL and 86 fg/mL for DPV and SWV, respectively, whereas the linear range was 1–300 pg/mL.

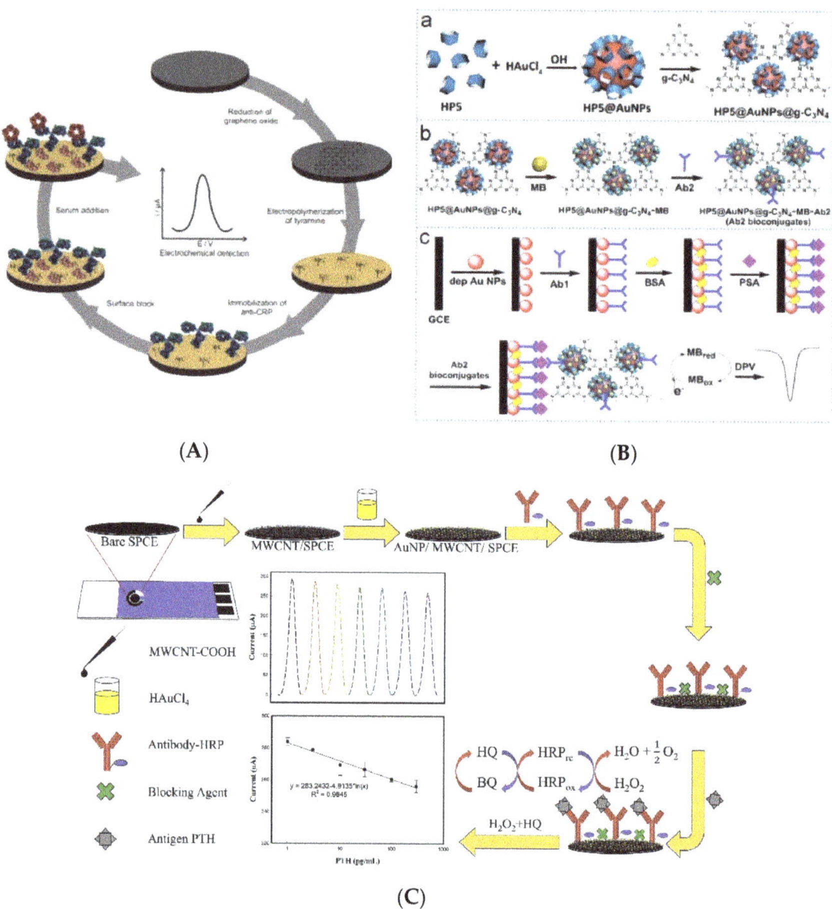

Figure 1. Examples of voltammetric immunosensors with non-covalently immobilized antibodies. (**A**) Schematic diagram of CRP immunosensor. Antibodies were immobilized on the polymer-modified electrode by the charge interaction. Reproduced with permission from [30]. Copyright (2020) John Wiley and Sons., Inc. (**B**) Immunosensor with non-covalently immobilized antibodies on AuNPs. Reproduced with permission from [33]. Copyright (2018) Elsevier. (**C**) Schematic diagram of immunosensor using gold and carbon-based materials at the same time. Reproduced with permission from [41]. Copyright (2021) Elsevier.

2.2. Chemical Immobilization of Antibodies

The chemical immobilization of antibodies using primary bonds has the advantage of providing robust, stable, and irreversible immobilization [2,42,43], which has led to its widespread use for the fabrication of immunoaffinity layers [42]. Shamsipur et al. developed an immunosensor based on the cross-linking of antibodies with silanized magnetite (Fe_3O_4) NPs to detect HER2 [43]. Magnetite NPs were functionalized with amine groups by treatment with APTMS, and antibodies were immobilized on the NPs by cross-linking with glutaraldehyde. Antibody-conjugated particles were immobilized on the electrode. The fabricated biosensor showed an LOD of 2.0×10^{-5} ng/mL with a linear range of 5.0×10^{-4}–50.0 ng/mL, based on the DPV measurement.

Similar to non-covalent immobilization, carbon-based materials have been widely used to chemically immobilize immunoaffinity layers. Rauf et al. utilized carboxyl-functionalized GO (GO–COOH) to fabricate a Mucin1 immunosensor [44]. As shown in Figure 2a, GO–COOH was assembled on the electrode and antibodies were immobilized using EDC. After Mucin1 was bound to the immunoaffinity layer, the signal was measured using DPV, and the LOD and linear range were calculated to be 0.04 U/mL and 0.1–50 U/mL, respectively. Devi et al. developed a microfluidic CysC immunosensor based on a CS–GO nanocomposite [45]. The electrode was modified using a nanocomposite and the antibodies against CysC were immobilized using EDC/NHS chemistry. The fabricated biosensor showed an LOD of 0.0078 mg/mL with a detection range of 1–10 mg/mL. Kalyani et al. fabricated a bio-nanocomposite using MWCNTs and magnetite NPs in CS, and used it to develop a CA19-9 immunosensor [46]. After depositing the bio-nanocomposite on the electrode, the antibodies were chemically immobilized using glutaraldehyde cross-linking (Figure 2b). The LOD and linear range were calculated to be 0.163 pg/mL and 0.001–100 ng/mL, respectively, based on SWV measurements.

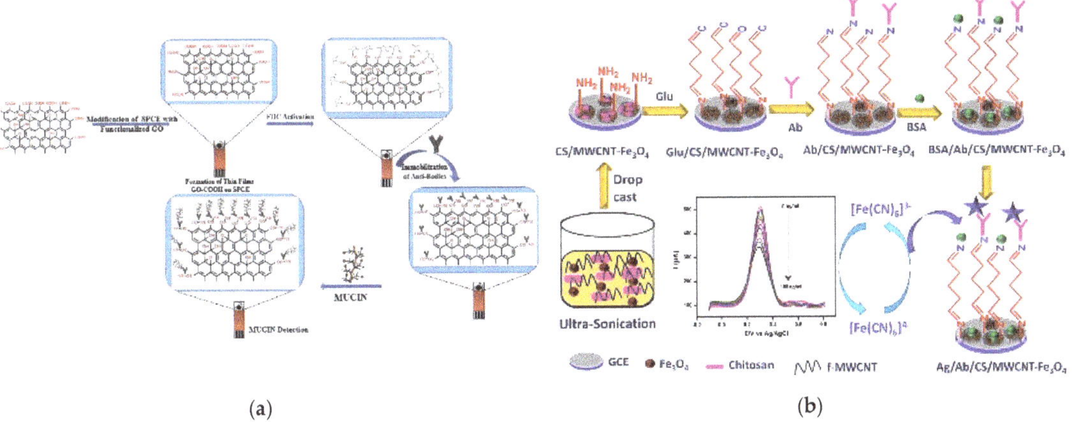

Figure 2. Examples of voltammetric immunosensors with covalently immobilized antibodies. (**a**) Immunosensor based on carboxyl functionalized GO. Antibodies were chemically immobilized using EDC chemistry. Reproduced with permission from [44]. Copyright (2018) Elsevier. (**b**) CA19-9 biosensor based on CS-MWCNT-Fe_3O_4. Antibodies were immobilized using glutaraldehyde. Reproduced with permission from [46]. Copyright (2021) Elsevier.

2.3. Other Methods

In the above section, immunosensors with antibodies immobilized using physical adsorption or chemical bonds were introduced. Unlike the studies described above, Sun et al. developed a voltammetric HSP70 immunosensor based on antigen immobilization [47]. They coated porous graphene onto the electrode and HSP70 was immobilized using physical adsorption. Then, the analyte HSP70 was pre-treated with biotinylated antibodies for a competitive assay. After the binding of the antibodies, streptavidin-conjugated HRP was added to generate the signal, which was measured using DPV. The fabricated immunosensor showed an LOD of 0.02 ng/mL with a linear range of 0.0448–100 ng/mL. Song et al. utilized affinity binding for the immobilization of antibodies to fabricate an electrochemical sensing system [48]. They immobilized avidin onto the ITO electrode using physical adsorption, and biotinylated capture antibodies were immobilized by biotin–avidin affinity binding. For signal generation, ALP-labeled detection antibodies were added after analyte binding. The fabricated immunosensor was tested to detect MMP-9 and Apo-A4 based on CV measurements. The developed immunosensor showed LODs of 0.21 and 6.6 ng/mL, with detection ranges 0.4–100 and 10–100 ng/mL for MMP-9 and Apo-A4 detection, respectively.

3. Amperometric Immunosensors

In amperometric immunosensors, the density or magnitude of the current is determined by measuring the electrochemical reactions at a constant voltage [9]. This technique has similar biosensing characteristics to those of other methods, such as the response time, dynamic range, and sensitivity [49,50]. In amperometry, reducing or oxidizing potential is generally applied to the working electrode, and the concentration of the reduced or oxidized substances is proportional to the measured current [51]. Similar to voltammetry, amperometry is one of the most widely used electrochemical analytical methods for immunosensors owing to its high selectivity. A high selectivity is achieved because the potential applied in the amperometric method is the specific reducing or oxidizing potential of the target [52]. In addition, amperometric biosensors require a minimal amount of analyte, so amperometric biosensors are suitable for monitoring analytes. Similar to the voltammetric immunosensors, a signal generation molecule is required in amperometric immunosensors, so the development of highly sensitive signal-generating or -amplifying composites is a key challenge for the fabrication of amperometric immunosensors.

3.1. Physical Adsorption of Antibodies

The physical adsorption of antibodies is also one of the most frequently used techniques for forming immunoaffinity layers. Chutichetpong et al. fabricated a disposable immunosensor based on a sandwich immunocomplex to detect MPT64 [53]. Capture antibodies were physically adsorbed onto the electrode to form an immunoaffinity layer, and HRP-conjugated detection antibodies were applied after antigen binding (Figure 3a). The signal was generated using the TMB reaction with HRP and measured using chronoamperometry. The fabricated immunosensor showed an LOD of 0.43 ng/mL, with two linear ranges of 0.3–50 ng/mL and 50–1000 ng/mL with different slopes.

Various metal composite-based sensors have been studied for the fabrication of amperometric immunosensors. Yan et al. developed a label-free amperometric biosensor for HBsAg detection [54]. In that study, AuPdCu ternary NPs were hydrothermally synthesized on N-GQDs, and the synthesized NPs were immobilized on PEI spheres using electrostatic attraction. The fabricated nanocomposites were immobilized on the electrode, whereas anti-HBsAg was immobilized using physical adsorption. After antigen binding, the signal was measured using amperometry, and the LOD and linear range were calculated to be 3.3 fg/mL and 10 fg/mL–50 ng/mL, respectively. Li et al. fabricated an amperometric immunosensor based on PdAg mesoporous nanospheres for the detection of multiple tumor markers [55]. As shown in Figure 3b, PdAg nanospheres were immobilized on the electrode and two antibodies against CEA and AFP were immobilized using physical adsorption. For signal generation, two types of detection antibodies were used: anti-CEA antibodies conjugated with $PdAgCeO_2$ mesoporous nanospheres and anti-AFP antibodies labeled with manganese dioxide (MnO_2) nanosheets. After sandwich immunocomplex formation, the first signal from the CEA and AFP was measured using amperometry. Then, the MnO_2 nanosheet was eliminated using acid treatment, and the second signal was measured. The concentrations of AFP and CES antigens calculated from the signal difference between the two measurements were 0.001 ng/mL and 0.0005 ng/mL, with linear ranges of 0.005–100 ng/mL and 0.001–40 ng/mL, respectively.

Gold nanomaterials have also been widely used as amperometric immunosensors. Zhang et al. developed an AFP biosensor based on antibodies physically adsorbed on AuNPs [56]. AuNPs were coated on the electrode using electrodeposition and anti-AFP antibodies were immobilized. For signal amplification, PDA-functionalized phenolic resin microporous carbon spheres were decorated with silver-coated NPs for detecting antibody conjugation. The signal generated by the reduction of H_2O_2 was measured using amperometry, and the LOD and linear range were calculated to be 6.7 fg/mL and 20fg/mL–100 ng/mL, respectively. Zhang et al. fabricated an immunoaffinity layer using AuNPs for the detection of CEA [57]. As shown in Figure 3c, an immunoaffinity layer was formed by immobilizing antibodies on the AuNP-coated electrode, and then microporous carbon spheres loaded with AgNP-spaced Hemin/rGO porous composite materials were used as a label for the detection antibodies. This composite improves the catalytic activity by reducing H_2O_2; thus, the amperometric signal was amplified by composite detection antibody-binding. The developed immunosensor showed an LOD of 6.7 fg/mL with a linear range of 20–200 ng/mL. Yola et al. developed an amperometric galectin-3 immunosensor [58]. In that study, the capture antibody was conjugated with AuNP-functionalized g-C_3N_4, and a Ti-based MOF coated with a covalent organic framework was used as a label for the detection antibody for signal amplification. After forming the sandwich immunocomplex, the amperometric signal was measured and the results showed LOQ and LOD values of 0.10 pg/mL and 0.025 pg/mL, respectively, with a 0.1 pg/mL–20 ng/mL linear range. Yan et al. fabricated immunosensors using AuNRs and GO to detect HE4Ag [59]. They conjugated titanium oxide nanocluster-functionalized nitrogen-doped rGO with Pd-functionalized AuNR (AuNR@Pd) using silanization and modified electrodes with conjugated composites. The antibodies against HE4 were then immobilized using physical adsorption. The developed immunosensor showed an LOD of 13.33 fM and a 40–60 nM linear range.

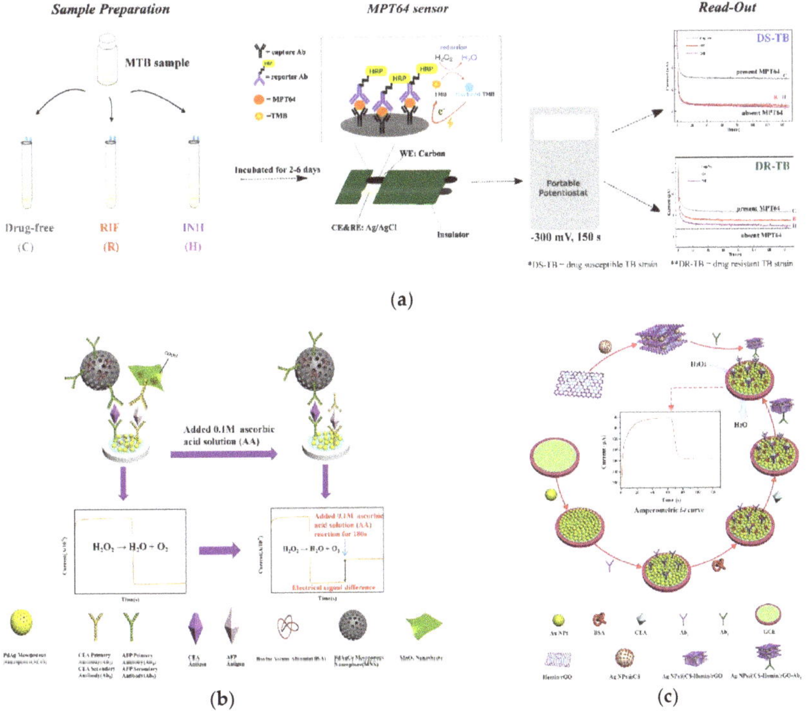

Figure 3. Examples of amperometric immunosensors with non-covalently immobilized antibodies. (**a**) Sandwich amperometric immunosensor. The immunoaffinity layer was fabricated by the direct treatment of antibodies on the electrode. Reproduced with permission [53]. Copyright (2018) Elsevier. (**b**) Composites-based immunosensor. Antibodies were immobilized on the PdAg nanosphere-modified electrode by physical adsorption. Reproduced with permission [55]. Copyright (2020) Elsevier. (**c**) AuNP-based immunosensor. An immunoaffinity layer was formed by immobilizing antibodies on the AuNP-coated electrode. Reproduced with permission [57]. Copyright (2020) Elsevier.

3.2. Chemical Immobilization of Antibodies

Various studies have used chemically immobilized antibodies for amperometric biosensors. Martínez-Periñán et al. fabricated an endoglin immunosensor based on chemically immobilized antibodies [60]. For effective antibody immobilization, the electrode was modified with pPPA using electropolymerization to obtain an abundance of carboxyl groups. Then, the capture antibodies were immobilized onto the carboxyl group using EDC/NHS. After the analyte was bound, biotinylated detection antibodies and poly-HRP-conjugated streptavidin were sequentially added to amplify the signal. This biosensor showed an LOD of 140 pg/mL with a linear range of 0.1–600 ng/mL. Ehzari et al. developed an amperometric immunosensor using magnetite NPs and MWCNTs to detect HER2 [61]. The electrode was sequentially modified with TMU-21-decorated Fe_3O_4 (Fe_3O_4@TMU-21) and carboxylated MWCNTs, and the capture antibodies were immobilized through EDC/NHS chemistry (Figure 4a). The LOD and linear range of the fabricated amperometric immunosensor were 0.3 pg/mL and 1.0 pg/mL–100 ng/mL, respectively.

AuNPs are also widely used for the chemical immobilization of antibodies. Hou et al. immobilized capture antibodies using EDC/NHS chemistry on AuNPs to detect EV71 [62]. The electrode was modified with AuNPs using electrochemical deposition, and the AuNPs were modified with carboxyl groups by SAMs. Next, antibodies against EV71 were chemically immobilized onto the electrode. Dual-labeled magnetic nanobeads with antibodies and HRP were used for signal amplification, and the signal from the TMB reaction was measured using amperometry. The fabricated immunosensor showed an LOD of 0.01 ng/mL with a 0.1–600 ng/mL linear range. Razzino et al. also used AuNPs to develop an amperometric immunosensor for the detection of tau protein [63]. As shown in Figure 4b, the AuNP-PAMAM dendrimer nanocomposite was chemically immobilized on the electrode. Capture antibodies were immobilized on the maximum amine group of the PAMAM dendrimer using glutaraldehyde cross-linking. The signal was amplified using HRP-labeled detection antibodies and the amperometric measurement showed an LOD of 1.7 pg/mL with a 6–5000 pg/mL linear range.

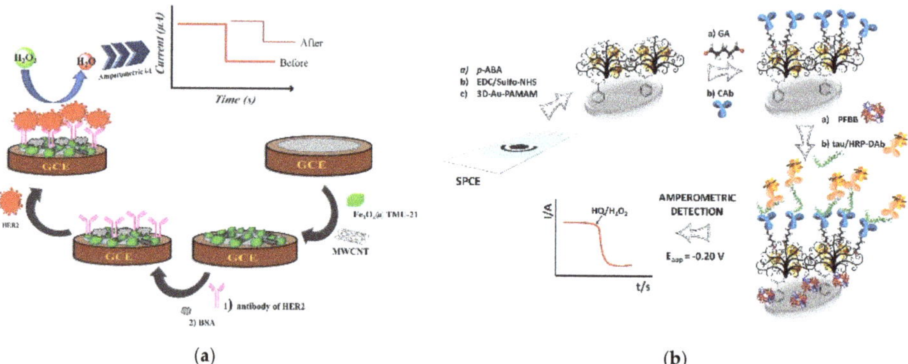

(a) (b)

Figure 4. Examples of amperometric immunosensors with covalently immobilized antibodies. (**a**) Magnetite nanoparticles and MWCNT-based immunosensor. Antibodies were chemically immobilized with carboxylated MWCNTs through EDC/NHS chemistry. Reproduced with permission [61]. Copyright (2020) Elsevier. (**b**) AuNP-based immunosensor. Antibodies were immobilized on the amine group of the PAuNP-AMAM dendrimer nanocomposite using glutaraldehyde cross-linking. Reproduced with permission [63]. Copyright (2020) Elsevier.

4. Impedimetric Immunosensors

Electrochemical EIS is an electrochemical analytical method that analyzes combined resistive and capacitive properties [51]. EIS measures electron and mass transfer by scanning alternating current frequencies and quantifies dielectric properties, including resistive and capacitive properties and impedance [64]. Impedance is usually expressed as a complex number consisting of a real component (ohmic resistance) and an imaginary component (capacitive reactance), and Nyquist and Bode plots are generally used to analyze electrochemical impedance data [65–67]. In comparison with the voltammetry and amperometry, an impedance biosensor has the strong advantage of not requiring labeling [68]. This means that impedimetric immunosensors can detect the binding of an analyte without any additional signal generation or amplifying molecules. In addition, impedance analysis is highly sensitive with a high degree of accurate responses, as well as being stable and reproducible. Therefore, impedance has been used in both biosensors and various other applications, such as for clinical diagnosis, food analysis, environmental monitoring, and battery analysis [69–71]. For impedimetric immunosensors, capturing antibodies that form an immunoaffinity layer is one of the most important factors; therefore, effective antibody immobilization on the surface of the electrode is directly linked to improved biosensor performance [72].

4.1. Physical Adsorption of Antibodies

Han et al. developed an impedimetric immunosensor using AuNR for the detection of *Staphylococcus aureus* [73]. They immobilized AuNRs onto an electrode using charge interaction. Before AuNR immobilization, positively charged poly-(diallyldimethylammonium chloride) and PSS were sequentially added to the electrode, and then positively charged AuNR was immobilized (Figure 5a). After AuNR immobilization, antibodies against *S. aureus* were immobilized using physical adsorption, and after bacterial binding, the signals were measured using EIS. The developed biosensor showed a 2.4×10^2 CFU/mL LOD with a 1.8×10^3–1.8×10^7 CFU/mL linear range.

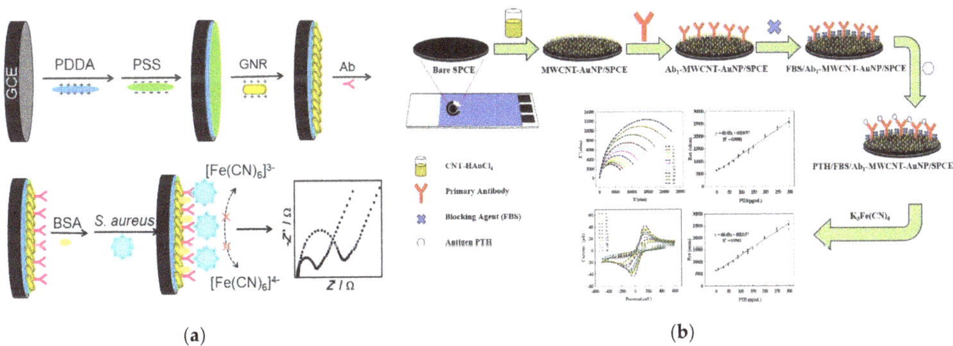

Figure 5. Examples of impedimetric immunosensors with non-covalently immobilized antibodies. (**a**) AuNR-based immunosensor. Antibodies were immobilized on the AuNR-modified electrode by physical adsorption. Reproduced with permission [73]. Copyright (2020) Elsevier. (**b**) MWCNT–AuNP nanocomposite-based immunosensor. The capture antibodies were immobilized on MWCNT–AuNP nanocomposite using physical adsorption. Reproduced with permission [74]. Copyright (2021) Elsevier.

Malla et al. fabricated a PTH immunosensor based on nanocomposites of AuNPs and MWCNTs [74]. As shown in Figure 5b, the synthesis of AuNPs on MWCNTs and the modification of the electrode with MWCNT–AuNP nanocomposites were conducted in a single step using controlled-potential electrodeposition. After the modification of the electrode, the anti-PTH antibodies were immobilized using physical adsorption, followed by hormone treatment. Then, the signals were measured using both CV and EIS, and the LOD of each measurement was calculated to be 0.092 pg/mL and 0.033 pg/mL, respectively, with a linear range of 1–300 pg/mL.

4.2. Chemical Immobilization of Antibodies

Various impedimetric immunosensors based on covalently immobilized antibodies have been developed, such as that of Nawaz et al., who immobilized antibodies on electrografted proteins through covalent bonds for the diagnosis of dengue virus [75]. BSA was electrografted onto the electrode to induce antifouling properties and enhanced conductivity, and anti-NS1 antibodies were chemically immobilized using EDC/NHS chemistry. The formation of an immunocomplex was confirmed using EIS measurement, and the calculated LOD was 0.3 ng/mL with a 1–200 ng/mL linear range. Nessark et al. developed an impedance immunosensor by electrografting a polymer to detect IL-10 [76]. For the fabrication of the transducer, silicon dioxide and silicon nitride were sequentially deposited onto a silicon substrate using chemical vapor deposition. Then, the immunoaffinity layer was fabricated using silanization and polymerization. The electrode was silanized using SPy and modified with PPy. Then, the polymer-modified electrode was electrochemically modified with a carboxyl group. Antibodies were immobilized on the electrode through the carboxyl group using EDC/NHS. The fabricated immunosensor showed a sensitivity of 0.1128 pg/mL and an LOD of 0.347 pg/mL, with a detection range of 1–50 pg/mL.

Sadighbayan et al. and Aydın et al. fabricated various impedimetric immunosensors with chemically immobilized antibody layers based on an ITO transducer [77–82]. For IL-8 detection, antibodies were immobilized onto a PHA SAM layer using EDC/NHS chemistry (LOD: 7.5 fg/mL) and an epoxy group was placed on PGMA and the CB-functionalized electrode (LOD: 3.3 fg/mL) [77,78]. To detect tumor marker p53, a spin-coated CS–CB composite layer and glutaraldehyde cross-linking were used (LOD: 3 fg/mL) with an epoxy group on the PGMA spin electrode (LOD: 3.3 fg/mL) [79,80]. For CCR4 detection, an ITO electrode with acid-substituted PPy (PPy-COOH) was used to maximize the antibody binding sites, and antibodies were chemically bound using EDC/NHS (LOD: 6.4 fg/mL) [81]. Recently, a SARS-CoV-2 impedimetric immunosensor was developed using AuNPs [82]. The AuNPs were capped with a SAM layer to expose the carboxyl group, and the modified NPs were immobilized onto the ITO electrode. Then, the antibodies were immobilized through chemical bonds using EDC/NHS, and the fabricated biosensor showed an LOD of 0.577 fg/mL with a 0.002–100 pg/mL linear range.

Carbon-based materials, such as MWCNTs, have also been used to fabricate impedimetric immunosensors. Simão et al. fabricated an MWCNT–AuNP-based immunoaffinity layer to detect ALP [83]. As shown in Figure 6, the MWCNT–AuNPs on the electrode were functionalized with carboxyl groups using cysteamine, and the antibodies were covalently immobilized using EDC/NHS chemistry. After analyte binding, the signal was measured using EIS, and the LOD was calculated to be 0.25 IU/L with a 0.5–600 IU/L detection range. Vasantham et al. fabricated a paper-based immunosensor utilizing carboxyl-functionalized MWCNTs to detect cTnI [84]. After covalently immobilizing the capture antibodies, the EIS measurement showed a 0.05 ng/mL LOD and 1.85 mΩ/ng/mL sensitivity, with a 0.05–50 ng/mL detection range.

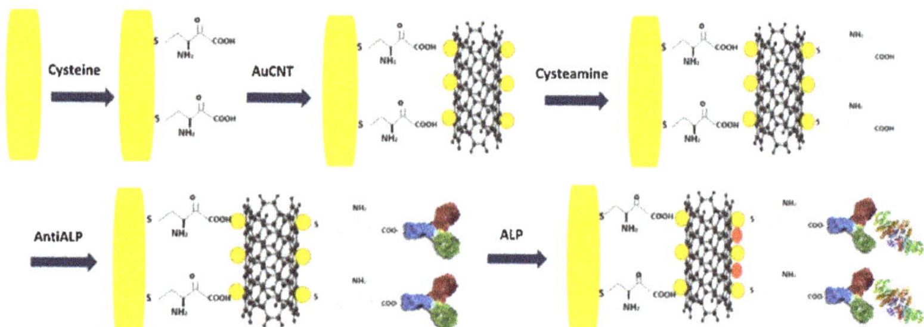

Figure 6. Example of impedimetric immunosensors with covalently immobilized antibodies. MWCNT–AuNP-based immunosensor was fabricated by the immobilization of antibodies on carboxyl-functionalized MWCNT–AuNP using EDC/NHS chemistry. Reproduced with permission [83]. Copyright (2018) Elsevier.

5. ECL Immunosensors

5.1. Physical Adsorption of Antibodies

ECL, also known as electrogenerated chemiluminescence, is a type of chemiluminescence in which luminophores generate electronically excited states by electron transfer on the electrode surface [85,86]. Consequently, the ECL technique does not require any external light source for excitation; therefore, it not only simplifies the measuring system, but it also reduces the background noise signal from the scattered light source or the autofluorescence of the analyte compared with a chemiluminescence system [87–89]. In addition, the ECL technique has become a powerful analytical method owing to its high sensitivity and stability [90–93]. Thus, ECL has been used for various applications, such as clinical diagnosis, environmental monitoring, and food monitoring [94–96]. However, ECL has the disadvantage of requiring instruments, including emission detection systems [97,98].

Liu et al. used AgNPs to form an immunoaffinity layer to detect cyclin D1 [99]. The electrode was modified with a PDA–AgNP composite, and capture antibodies were immobilized using physical adsorption. For signal generation, detection antibodies conjugated with AuNPs and Bi_2S_3 QD-based nanoprobes were used. After forming the sandwich immunocomplex, the ECL measurement showed a 6.34 fg/mL LOD with a 10 fg/mL–1 μg/mL linear range. Du et al. fabricated an ECL immunosensor with CdS QDs and AgNPs for the detection of cTn-I [100]. In that study, the luminophore was firstly immobilized on an immunoaffinity layer (Figure 7). The synthesized nanoluminophore, MOF-5-encapsulated CdS QD (CdSQD@MOF-5), was coated with a PDDA-modified electrode. Then, AgNPs-conjugated antibodies were immobilized to form an immunoaffinity layer. cTnI was quantified using an ECL-based measurement and the result showed a 5.01 fg/mL LOD with a 0.01–1000 pg/mL linear range.

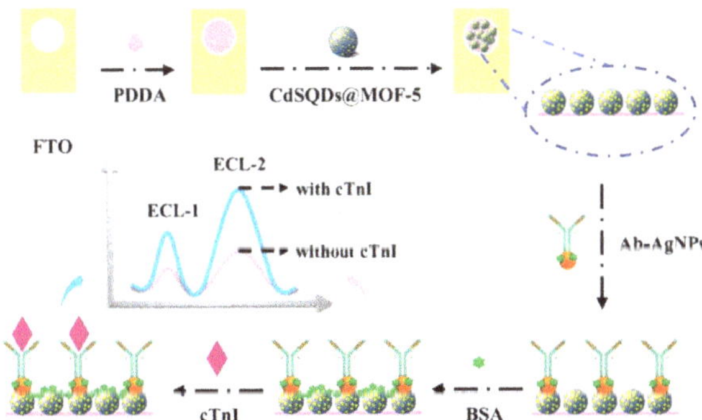

Figure 7. Examples of ECL immunosensors with physically adsorbed capture antibodies. Antibodies were immobilized on AgNPs to form an immunoaffinity layer. Reproduced with permission [100]. Copyright (2020) American Chemical Society.

Various ECL immunosensors have been developed using an AuNP-based immunoaffinity layer with physically adsorbed capture antibodies. Studies by Wang et al. and Yang et al. from the Yuan group developed ECL immunosensors using physically immobilized antibodies on electrodeposited AuNPs [101,102]. They used a compound consisting of RU derivative (RUD), PEI, and ABEI as luminophores to increase the ECL signal using ECL resonance energy transfer to detect Col IV [101]. This immunosensor showed a 0.17 pg/mL LOD with a 0.5 pg/mL–7.2 ng/mL detection range. In another study, they used PFO dots instead of RUD to fabricate a KIM-1 immunosensor [102]. The ECL emitter, an ABEI-PER-PFO dot, was combined with rGO-PtNP for conjugation with detection antibodies. After forming the sandwich immunocomplex, the LOD and linear range were calculated to be 16.7 fg/mL and 50 fg/mL–1 ng/mL, respectively, using ECL measurement. Zheng et al. fabricated a CEA immunosensor using an AuNP–Ab immunoaffinity layer [103].

Antibodies were immobilized onto the AuNP-modified electrode using physical adsorption. To generate ECL signals, compounds consisting of PDDA-rGO and ZnSe/ZnS QDs were conjugated with detection antibodies. CEA levels were then measured using ECL, and the fabricated immunosensor showed an LOD of 0.029 pg/mL with a linear range of 0.0001–100 ng/mL. Lian et al. developed a label-free ECL immunosensor using Au–Co alloy NPs (Au–Co NPs) for the detection of LDL and oxidized LDL (ox-LDL) [104]. Au–Co NPs were immobilized onto an APTMS-silanized electrode, and each antibody against LDL and ox-LDL was immobilized. After antigen binding and ECL measurement, the calculated LODs were 0.256 pg/mL and 0.330 pg/mL, and the calculated linear ranges were 0.420–100 pg/mL and 0.500–60.0 pg/mL, for LDL and ox-LDL, respectively.

Gold nanomaterials have also been used with carbon-based materials for the fabrication of immunoaffinity layers. Qin et al. modified an electrode with a nanocomposite consisting of Ru–SiO_2, AuNP, and rGO (rGO@AuNP@RU-SiO_2) to detect AFP using ECL [105]. After sequential antibody and antigen treatment, the ECL measurement showed a 0.03 pg/mL LOD, with a 0.0001–100 ng/mL linear range. Khan et al. fabricated a label-free ECL immunosensor using $Ce_2Sn_2O_7$–AuNP nanocubes for the detection of CEA [106]. For the fabrication of the immunosensor, nanocubes and capture antibodies were sequentially treated. After antigen treatment, the ECL signal was measured, and the developed immunosensor presented a linear range of 0.001–70 ng/mL and an LOD of 0.53 pg/mL.

5.2. *Chemical Immobilization of Antibodies*

Covalent immobilization is also frequently used for the formation of immunoaffinity layers in ECL biosensors. Babamiri et al. fabricated an HBsAg immunosensor using magnetite NPs [107]. Antibodies against HBsAg were covalently immobilized onto carboxyl-modified Fe_3O_4 NPs via EDC/NHS chemistry. For the amplification of the ECL signal, CdTe/CdS QDs were used to form nanoclusters with PAMAM dendrimers, and detection antibodies were immobilized onto the fabricated nanoclusters. After the formation of the sandwich immunocomplex, HBsAg was quantified using ECL measurement, and the results showed a 0.80 fg/mL LOD and a 3 fg/mL–0.3 ng/mL linear range.

Fang et al. developed an ECL immunosensor using a carbon-based nanomaterial, g-C_3N_4, to detect HE4 [108]. They synthesized carboxyl-wrapped g-C_3N_4 and made it into a composite with mesoporous silica (g-C_3N_4@SiO_2). The capture antibodies were immobilized via the carboxyl group of g-C_3N_4 by EDC/NHS. For signal amplification, detection antibodies were chemically immobilized onto the carbon nanohorn–polymer dot composites. The fabricated ECL immunosensor showed a 3.3×10^{-6} ng/mL LOD with a 1.0×10^{-5}–10 ng/mL linear range. Ding et al. used another carbon-based nanomaterial, MWCNT, to detect 5hmC [109]. For the immobilization of antibodies, the electrode was modified with MWCNTs, and capture antibodies were chemically immobilized using EDC/NHS chemistry. After analyte treatment, detection antibodies conjugated with PAMAM–silver nanoclusters and nitrogen-doped graphene nanocomposites were used as ECL probes, and signals were measured using ECL. The calculated LOD and linear range were 2.47 pM and 10 pM–30 nM, respectively.

Yang et al. used AuNPs for the detection of PSA [110]. In that study, PICA was electropolymerized onto electrodeposited flower-like AuNPs to form PICA–AuNP nanocomposites (Figure 8). The capture antibodies were then chemically immobilized using EDC/NHS. For the ECL probe, detection antibodies were immobilized onto GQDs and AuNP-immobilized PEI-modified GO. The fabricated ECL immunosensor showed a 0.44 pg/mL LOD with a 0.001–10 ng/mL linear range.

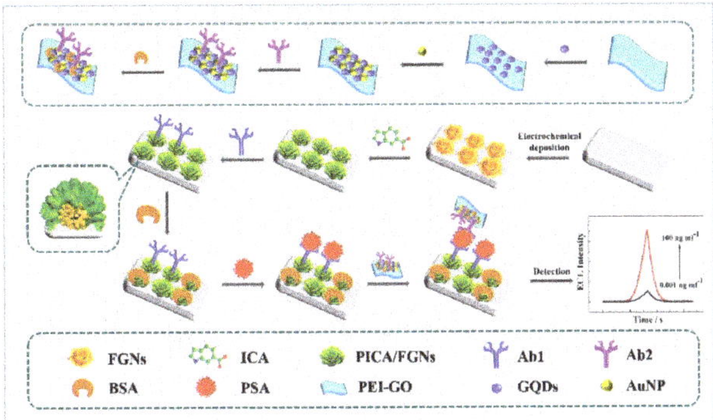

Figure 8. Examples of ECL immunosensors with covalently immobilized capture antibodies. The capture antibodies were chemically immobilized on a PICA–AuNP nanocomposite-modified electrode using EDC/NHS. Reproduced with permission [110]. Copyright (2020) Elsevier.

6. Conclusions and Perspectives

In this review, various electrochemical, voltammetric, amperometric, impedimetric, and ECL immunosensors have been introduced, as summarized in Table 1 and further discussed. In particular, we focused on the immobilization of antibodies for the fabrication of immunoaffinity layers, by dividing them into those immobilized using physical adsorption and those using chemical bonding. Generally, physical adsorption methods use not only simple immersion, but also attractive materials known to have an affinity for antibodies, such as AuNPs. To improve the performance of immunosensors, enhancing the effectiveness of antibodies, especially by increasing exposed antigen binding sites, is a key technique. Therefore, advanced antibody adsorption strategies, such as introducing the orientation control of antibodies, scFv, or nanobodies, would be advantageous.

Table 1. Examples of electrochemical immunosensors.

Methods	Immunoaffinity Layer	Analyte	LOD	Detection Range	Ref.
DPV	GO/tryamine/Ab [1]	CRP	1.25 μg/L (LOQ)	1.09–100 μg/L	[30]
	AuNP/Ab	PSA	0.12 pg/mL	0.0005–10.00 ng/mL	[33]
	Graphene/PPD/AuNP/Ab	NSE	0.3 ng/mL	1–1000 ng/mL	[37]
	CoS_2-graphene/AuNP/Ab	CA15-3	0.03 U/mL	0.1–150 U/mL	[38]
	Graphene/MOFs/Ab	NMP22	1.7 pg/mL	0.005–20 ng/mL	[39]
	MWCNT/PANI/AuNP/Ab	PSA	0.5 pg/mL	1.66 ag/mL–1.3 ng/mL	[40]
	APTMS-Fe_3O_4/Ab	HER2	20 fg/mL	5×10^{-4}–50 ng/mL	[43]
	GO–COOH/Ab	Mucin1	0.04 U/mL	0.1–2 U/mL	[44]
	GO–CS/Ab	CysC	0.0078 mg/mL	1–10 mg/mL	[45]
	Graphene/Antigen	HSP70	0.02 ng/mL	0.0448–100 ng/mL	[47]
DPV SWV	MWCNT/AuNP/Ab	PTH	886 fg/mL 65 fg/mL	1–300 pg/mL	[41]
SWV	AF-GO/PEI/Ab	CysC	0.03 ng/mL	0.1–1000 ng/mL	[31]
	CS/AuNP/Ab	PSA	0.001 ng/mL	1–18 ng/mL	[34]
	PANI/AuNP/Ab	PSA	1.25 fg/mL	0.01–100 ng/mL	[35]
	CS-MWCNT-Fe_3O_4/Ab	CA19-9	0.163 pg/mL	0.001–100 ng/mL	[46]
SWV Amperometry	Au/Ab	procalcitonin	82.6 fg/mL 95.4 fg/mL	0.0001–50 ng/mL	[36]
CV	Avidin/biotinylated Ab	MMP-9 Apo-A4	0.21 ng/mL 6.6 ng/mL	0.4–100 ng/mL 10–100 ng/mL	[48]

Table 1. *Cont.*

Methods	Immunoaffinity Layer	Analyte	LOD	Detection Range	Ref.
Amperometry	Ab	MPT64	0.43 ng/mL	0.3–50 ng/mL 50–1000 ng/mL	[53]
	N-GQD/AuPdCu/Ab	HBsAg	3.3 fg/mL	10 fg/mL–50 ng/mL	[54]
	PtAg/Ab	CEA AFP	0.0005 ng/mL 0.001 ng/mL	0.001–40 ng/mL 0.005–100 ng/mL	[55]
	AuNP/Ab	AFP	6.7 fg/mL	20 fg/mL–100 ng/mL	[56]
	AuNP/Ab	CEA	6.7 fg/mL	20 fg/mL–200 ng/mL	[57]
	g-C_3N_4/AuNP/Ab	galectin-3	25.0 fg/mL	0.0001–20.0 ng/mL	[58]
	TiO_2-rGO/AuNR@Pd/Ab	HE4Ag	13.33 fM	40 fM–60 nM	[59]
	pPPA/Ab	Endoglin	140 pg/mL	0.18–20 ng/mL	[60]
	Fe_3O_4@TMU-21/MWCNT/Ab	HER2	0.3 pg/mL	1.0 pg/mL–100 ng/mL	[61]
	AuNP/Ab	EV71	0.01 ng/mL	0.1–600 ng/mL	[62]
	AuNP-PAMAM/Ab	tau	1.7 pg/mL	6–5000 pg/mL	[63]
EIS	PDDA/PSS/AuNR/Ab	*S. aureus*	2.4×10^2 CFU/mL	1.8×10^3–1.8×10^7 CFU/mL	[73]
	BSA/Ab	NS1	0.3 ng/mL	1–200 ng/mL	[75]
	SPy-PPy/Ab	IL-10	0.347 pg/mL	1–50 pg/mL	[76]
	PHA/Ab	IL-8	7.5 fg/mL	0.025–3 pg/mL	[77]
	PGMA-CB/Ab	IL-8	3.3 fg/mL	0.01–3 pg/mL	[78]
	Chitosan-CB	P53	3 fg/mL	0.01–2 pg/mL	[79]
	PGMA/Ab	P53	7 fg/mL	0.02–4 pg/mL	[80]
	PPy-COOH/Ab	CCR4	6.4 fg/mL	0.02–8 pg/mL	[81]
	AuNP-SAM/Ab	SARS-CoV-2	0.577 fg/mL	0.002–100 pg/mL	[82]
	MWCNT–AuNP/Ab	ALP	0.25 IU/L	0.5–600 IU/L	[83]
	MWCNT-COOH/Ab	cTnI	0.05 ng/mL	0.05–50 ng/mL	[84]
EIS CV	MWCNT–AuNP/Ab	PTH	0.033 pg/mL 0.092 pg/mL	1–300 pg/mL	[74]
ECL	PDA-AgNP/Ab	Cyclin D1	6.34 fg/mL	10 fg/mL–1 μg/mL	[99]
	CdSQD@MOF-5/AgNP-Ab	cTnI	5.01 fg/mL	0.01–1000 pg/mL	[100]
	AuNP/Ab	Col IV	0.17 pg/mL	0.5 pg/mL–7.2 ng/mL	[101]
	AuNP/Ab	KIM-1	16.7 fg/mL	50 fg/mL–1 ng/mL	[102]
	AuNP/Ab	CEA	0.029 pg/mL	0.0001–100 ng/mL	[103]
	APTMS/Au–Co NPs/Ab	LDL Ox-LDL	0.256 pg/mL 0.330 pg/mL	0.420–100 pg/mL 0.5–60 pg/mL	[104]
	rGO@AuNP@RU-SiO_2/Ab	AFP	0.03 pg/mL	0.0001–100 ng/mL	[105]
	$Ce_2Sn_2O_7$-AuNP/Ab	CEA	0.53 pg/mL	0.001–70 ng/mL	[106]
	Fe_3O_4/Ab	HBsAg	0.80 fg/mL	3 fg/mL–0.3 ng/mL	[107]
	g-C_3N_4@SiO_2/Ab	HE4	$3.3\times \times 10^{-6}$ ng/mL	$1.0\times \times 10^{-5}$–10 ng/mL	[108]
	MWCNT/Ab	5hmC	2.47 pM	10 pM–30 nM	[109]
	AuNP/PICA/Ab	PSA	0.44 pg/mL	0.001–100 ng/mL	[110]

[1] Antibodies.

For covalent immobilization, electrodes or binding materials are modified with functional groups such as carboxyl, amine, and hydroxyl using various chemistry-based techniques, including SAM and silanization. Among them, EDC/NHS chemistry is one of the most frequently used methods for binding the amine groups of antibodies to carboxylated surfaces. In addition, silanization methods are frequently used for hydroxyl surfaces, such as ITO electrodes. Covalent immobilization can irreversibly immobilize antibodies to form a robust immunoaffinity layer. However, most chemical reactions require the linking and blocking of the functional groups in antibodies, which can affect the binding activity. In addition, blocking the Fab region, including the antigen-binding paratope, would limit the performance of immunosensors. Thus, Fc-binding strategies can improve the performance of electrochemical immunosensors by enhancing the functionalization of electrodes with high density and efficiency.

Electrochemical biosensors have been developed in the past few decades, and remain active research areas in analytical chemistry that offer advantages such as small sample volume, cost-effectiveness, simplicity, high sensitivity, reproducibility, and selectivity. For electrochemical immunosensors, antibodies are used for the recognition and specific binding of analytes onto electrodes, and the labeling of bound target analytes with signal generating or amplifying molecules. In electrochemical biosensors, various nanomaterials are introduced into the electrodes or labeling molecules for various reasons, such as increasing conductivity, electron transfer, signal generation, and amplification. Therefore, the synthesis of highly effective nanomaterials would increase the performance of electrochemical biosensors. In conclusion, strategies for introducing highly effective nanomaterials or their composite conjugated antibodies with high-density antigen-binding sites need to be assessed in future studies.

Supplementary Materials: The following are available online at https://www.mdpi.com/article/10.3390/bios11100360/s1. Table S1: Abbreviations used in this work.

Author Contributions: Conceptualization, M.P.; writing, J.K. and M.P.; supervision, M.P. All authors have read and agreed to the published version of the manuscript.

Funding: This work was supported by a National Research Foundation of Korea (NRF) grant funded by the Korean Government (MSIP) (2021R1I1A305570011) and by the Technology Innovation Program (20008414, Development of intestine–liver–kidney multiorgan tissue chip mimicking absorption, distribution, metabolism, and excretion of drugs) funded by the Ministry of Trade, Industry and Energy (MOTIE, Korea), Republic of Korea.

Institutional Review Board Statement: Not applicable.

Informed Consent Statement: Not applicable.

Data Availability Statement: Not applicable.

Conflicts of Interest: The authors declare no conflict of interest.

References

1. Park, M.; Heo, Y.J. Biosensing technologies for chronic diseases. *BioChip J.* **2021**, *15*, 1–13. [CrossRef]
2. Park, M. Orientation control of the molecular recognition layer for improved sensitivity: A review. *BioChip J.* **2019**, *13*, 82–94. [CrossRef]
3. Tu, J.; Torrente-Rodríguez, R.M.; Wang, M.; Gao, W. The era of digital health: A review of portable and wearable affinity biosensors. *Adv. Funct. Mater.* **2020**, *30*, 1906713. [CrossRef]
4. Liu, D.; Wang, J.; Wu, L.; Huang, Y.; Zhang, Y.; Zhu, M.; Wang, Y.; Zhu, Z.; Yang, C. Trends in miniaturized biosensors for point-of-care testing. *TrAC Trends Anal. Chem.* **2020**, *122*, 115701. [CrossRef]
5. Mi, F.; Guan, M.; Hu, C.; Peng, F.; Sun, S.; Wang, X. Application of lectin-based biosensor technology in the detection of foodborne pathogenic bacteria: A review. *Analyst* **2021**, *146*, 429–443. [CrossRef]
6. Lee, D.; Park, K.; Seo, J. Recent advances in anti-inflammatory strategies for implantable biosensors and medical implants. *BioChip J.* **2020**, *14*, 48–62. [CrossRef]
7. Rodovalho, V.; Alves, L.; Castro, A.; Madurro, J.; Brito-Madurro, A.; Santos, A. Biosensors applied to diagnosis of infectious diseases—An update. *Austin J. Biosens. Bioelectron.* **2015**, *1*, 10–15.
8. Han, Y.D.; Chun, H.J.; Yoon, H.C. Low-cost point-of-care biosensors using common electronic components as transducers. *BioChip J.* **2020**, *14*, 32–47. [CrossRef]
9. Harshavardhan, S.; Rajadas, S.E.; Vijayakumar, K.K.; Durai, W.A.; Ramu, A.; Mariappan, R. Electrochemical immunosensors Working principle, types, scope, applications, and future prospects. In *Bioelectrochemical Interface Engineering*; Wiley: Hoboken, NJ, USA, 2019; pp. 343–369.
10. Park, M. Surface display technology for biosensor applications: A review. *Sensors* **2020**, *20*, 2775. [CrossRef]
11. Carrascosa, L.G.; Moreno, M.; Alvarez, M.; Lechuga, L.M. Nanomechanical biosensors: A new sensing tool. *TrAC Trends Anal. Chem.* **2006**, *25*, 196–206. [CrossRef]
12. Lazcka, O.; Del Campo, F.J.; Munoz, F.X. Pathogen detection: A perspective of traditional methods and biosensors. *Biosens. Bioelectron.* **2007**, *22*, 1205–1217. [CrossRef]
13. Deyev, S.; Lebedenko, E. Modern technologies for creating synthetic antibodies for clinical application. *Acta Nat.* **2009**, *1*, 32–51. [CrossRef]

14. Kim, J.K.; Song, H.-M.; Jun, J.W.; Park, H.J.; Lim, E.-S.; Lee, K.; Lee, S.; Kim, S. Clinical studies of Ci-5, Sol-gel encapsulated multiplex antibody microarray for quantitative fluorometric detection of simultaneous five different tumor antigens. *BioChip J.* **2019**, *13*, 378–386. [CrossRef]
15. Kang, J.; Kim, M.-G. Advancements in DNA-assisted Immunosensors. *BioChip J.* **2020**, *14*, 18–31. [CrossRef]
16. Hasan, A.; Nurunnabi, M.; Morshed, M.; Paul, A.; Polini, A.; Kuila, T.; Al Hariri, M.; Lee, Y.-K.; Jaffa, A.A. Recent advances in application of biosensors in tissue engineering. *BioMed Res. Int.* **2014**, *2014*, 307519. [CrossRef] [PubMed]
17. Thévenot, D.R.; Toth, K.; Durst, R.A.; Wilson, G.S. Electrochemical biosensors: Recommended definitions and classification. *Biosens. Bioelectron.* **2001**, *16*, 121–131. [CrossRef]
18. Hammond, J.L.; Formisano, N.; Estrela, P.; Carrara, S.; Tkac, J. Electrochemical biosensors and nanobiosensors. *Essays Biochem.* **2016**, *60*, 69–80.
19. Li, H.; Liu, X.; Li, L.; Mu, X.; Genov, R.; Mason, A.J. CMOS electrochemical instrumentation for biosensor microsystems: A review. *Sensors* **2017**, *17*, 74. [CrossRef]
20. Juska, V.B.; Pemble, M.E. A critical review of electrochemical glucose sensing: Evolution of biosensor platforms based on advanced nanosystems. *Sensors* **2020**, *20*, 6013. [CrossRef]
21. Zhang, S.; Wright, G.; Yang, Y. Materials and techniques for electrochemical biosensor design and construction. *Biosens. Bioelectron.* **2000**, *15*, 273–282. [CrossRef]
22. Schmidt-Speicher, L.M.; Länge, K. Microfluidic integration for electrochemical biosensor applications. *Curr. Opin. Electrochem.* **2021**, *29*, 100755. [CrossRef]
23. Porada, R.; Jedlińska, K.; Lipińska, J.; Baś, B. Voltammetric sensors with laterally placed working electrodes: A review. *J. Electrochem. Soc.* **2020**, *167*, 037536. [CrossRef]
24. Hussain, G.; Silvester, D.S. Comparison of voltammetric techniques for ammonia sensing in ionic liquids. *Electroanalysis* **2018**, *30*, 75–83. [CrossRef]
25. Amro, A.N. Voltammetric method development for itopride assay in a pharmaceutical formulation. *Curr. Pharm. Anal.* **2020**, *16*, 312–318. [CrossRef]
26. Park, M.; Kim, J.; Kim, K.; Pyun, J.C.; Sung, G.Y. Parylene-coated polytetrafluoroethylene-membrane-based portable urea sensor for real-time monitoring of urea in peritoneal dialysate. *Sensors* **2019**, *19*, 4560. [CrossRef] [PubMed]
27. Kim, J.Y.; Sung, G.Y.; Park, M. Efficient portable urea biosensor based on urease immobilized membrane for monitoring of physiological fluids. *Biomedicines* **2020**, *8*, 596. [CrossRef] [PubMed]
28. Pallela, R.; Chandra, P.; Noh, H.-B.; Shim, Y.-B. An amperometric nanobiosensor using a biocompatible conjugate for early detection of metastatic cancer cells in biological fluid. *Biosens. Bioelectron.* **2016**, *85*, 883–890. [CrossRef] [PubMed]
29. Mistry, K.K.; Layek, K.; Mahapatra, A.; RoyChaudhuri, C.; Saha, H. A review on amperometric-type immunosensors based on screen-printed electrodes. *Analyst* **2014**, *139*, 2289–2311. [CrossRef]
30. Ribeiro, S.H.; Alves, L.M.; Flauzino, J.M.; Moço, A.C.; Segatto, M.S.; Silva, J.P.; Borges, L.F.; Madurro, J.M.; Madurro, A.G. Reusable immunosensor for detection of C-reactive protein in human serum. *Electroanalysis* **2020**, *32*, 2316–2322. [CrossRef]
31. Trindade, E.K.; Silva, B.V.; Dutra, R.F. A probeless and label-free electrochemical immunosensor for cystatin C detection based on ferrocene functionalized-graphene platform. *Biosens. Bioelectron.* **2019**, *138*, 111311. [CrossRef]
32. Abad, J.M.; Puertas, S.; Pérez, D.; Sánchez-Espinel, C. Design and development of antibody functionalized gold nanoparticles for biomedical applications. *J. Nanosci. Nanotechnol.* **2021**, *21*, 2834–2840. [CrossRef]
33. Zhou, X.; Yang, L.; Tan, X.; Zhao, G.; Xie, X.; Du, G. A robust electrochemical immunosensor based on hydroxyl pillar [5] arene@ AuNPs@ g-C3N4 hybrid nanomaterial for ultrasensitive detection of prostate specific antigen. *Biosens. Bioelectron.* **2018**, *112*, 31–39. [CrossRef]
34. Suresh, L.; Brahman, P.K.; Reddy, K.R.; Bondili, J. Development of an electrochemical immunosensor based on gold nanoparticles incorporated chitosan biopolymer nanocomposite film for the detection of prostate cancer using PSA as biomarker. *Enzyme Microb. Technol.* **2018**, *112*, 43–51. [CrossRef]
35. Zhang, D.; Li, W.; Ma, Z. Improved sandwich-format electrochemical immunosensor based on "smart" SiO2@ polydopamine nanocarrier. *Biosens. Bioelectron.* **2018**, *109*, 171–176. [CrossRef] [PubMed]
36. Li, Y.; Liu, L.; Liu, X.; Ren, Y.; Xu, K.; Zhang, N.; Sun, X.; Yang, X.; Ren, X.; Wei, Q. A dual-mode PCT electrochemical immunosensor with CuCo2S4 bimetallic sulfides as enhancer. *Biosens. Bioelectron.* **2020**, *163*, 112280. [CrossRef]
37. Amani, J.; Maleki, M.; Khoshroo, A.; Sobhani-Nasab, A.; Rahimi-Nasrabadi, M. An electrochemical immunosensor based on poly p-phenylenediamine and graphene nanocomposite for detection of neuron-specific enolase via electrochemically amplified detection. *Anal. Biochem.* **2018**, *548*, 53–59. [CrossRef]
38. Khoshroo, A.; Mazloum-Ardakani, M.; Forat-Yazdi, M. Enhanced performance of label-free electrochemical immunosensor for carbohydrate antigen 15-3 based on catalytic activity of cobalt sulfide/graphene nanocomposite. *Sens. Actuators B Chem.* **2018**, *255*, 580–587. [CrossRef]
39. Zhao, S.; Zhang, Y.; Ding, S.; Fan, J.; Luo, Z.; Liu, K.; Shi, Q.; Liu, W.; Zang, G. A highly sensitive label-free electrochemical immunosensor based on AuNPs-PtNPs-MOFs for nuclear matrix protein 22 analysis in urine sample. *J. Electroanal. Chem.* **2019**, *834*, 33–42. [CrossRef]

40. Assari, P.; Rafati, A.A.; Feizollahi, A.; Joghani, R.A. Fabrication of a sensitive label free electrochemical immunosensor for detection of prostate specific antigen using functionalized multi-walled carbon nanotubes/polyaniline/AuNPs. *Mater. Sci. Eng. C* **2020**, *115*, 111066. [CrossRef] [PubMed]
41. Chen, G.-C.; Liu, C.-H.; Wu, W.-C. Electrochemical immunosensor for serum parathyroid hormone using voltammetric techniques and a portable simulator. *Anal. Chim. Acta* **2021**, *1143*, 84–92. [CrossRef]
42. Jung, Y.; Jeong, J.Y.; Chung, B.H. Recent advances in immobilization methods of antibodies on solid supports. *Analyst* **2008**, *133*, 697–701. [CrossRef] [PubMed]
43. Shamsipur, M.; Emami, M.; Farzin, L.; Saber, R. A sandwich-type electrochemical immunosensor based on in situ silver deposition for determination of serum level of HER2 in breast cancer patients. *Biosens. Bioelectron.* **2018**, *103*, 54–61. [CrossRef] [PubMed]
44. Rauf, S.; Mishra, G.K.; Azhar, J.; Mishra, R.K.; Goud, K.Y.; Nawaz, M.A.H.; Marty, J.L.; Hayat, A. Carboxylic group riched graphene oxide based disposable electrochemical immunosensor for cancer biomarker detection. *Anal. Biochem.* **2018**, *545*, 13–19. [CrossRef]
45. Devi, K.S.; Krishnan, U.M. Microfluidic electrochemical immunosensor for the determination of cystatin C in human serum. *Microchim. Acta* **2020**, *187*, 585. [CrossRef] [PubMed]
46. Kalyani, T.; Sangili, A.; Nanda, A.; Prakash, S.; Kaushik, A.; Jana, S.K. Bio-nanocomposite based highly sensitive and label-free electrochemical immunosensor for endometriosis diagnostics application. *Bioelectrochemistry* **2021**, *139*, 107740. [CrossRef] [PubMed]
47. Sun, B.; Cai, J.; Li, W.; Gou, X.; Gou, Y.; Li, D.; Hu, F. A novel electrochemical immunosensor based on PG for early screening of depression markers-heat shock protein 70. *Biosens. Bioelectron.* **2018**, *111*, 34–40. [CrossRef] [PubMed]
48. Song, S.; Kim, Y.J.; Shin, I.-S.; Kim, W.-H.; Lee, K.-N.; Seong, W.K. Electrochemical immunoassay based on indium tin oxide activity toward a alkaline phosphatase. *BioChip J.* **2019**, *13*, 387–393. [CrossRef]
49. Won, S.-Y.; Chandra, P.; Hee, T.S.; Shim, Y.-B. Simultaneous detection of antibacterial sulfonamides in a microfluidic device with amperometry. *Biosens. Bioelectron.* **2013**, *39*, 204–209. [CrossRef]
50. Chandra, P.; Noh, H.-B.; Won, M.-S.; Shim, Y.-B. Detection of daunomycin using phosphatidylserine and aptamer co-immobilized on Au nanoparticles deposited conducting polymer. *Biosens. Bioelectron.* **2011**, *26*, 4442–4449. [CrossRef]
51. Mahato, K.; Kumar, S.; Srivastava, A.; Maurya, P.K.; Singh, R.; Chandra, P. Electrochemical immunosensors. In *Handbook of Immunoassay Technologies*; Elsevier: Amsterdam, The Netherlands, 2018; pp. 359–414.
52. Lim, S.A.; Ahmed, M.U. Chapter 1—Introduction to immunosensors. In *Immunosensors*; The Royal Society of Chemistry: Cambridge, UK, 2019; pp. 1–20.
53. Chutichetpong, P.; Cheeveewattanagul, N.; Srilohasin, P.; Rijiravanich, P.; Chaiprasert, A.; Surareungchai, W. Rapid screening drug susceptibility test in tuberculosis using sandwich electrochemical immunosensor. *Anal. Chim. Acta* **2018**, *1025*, 108–117. [CrossRef]
54. Yan, Q.; Yang, Y.; Tan, Z.; Liu, Q.; Liu, H.; Wang, P.; Chen, L.; Zhang, D.; Li, Y.; Dong, Y. A label-free electrochemical immunosensor based on the novel signal amplification system of AuPdCu ternary nanoparticles functionalized polymer nanospheres. *Biosens. Bioelectron.* **2018**, *103*, 151–157. [CrossRef] [PubMed]
55. Li, W.; Yang, Y.; Ma, C.; Song, Y.; Qiao, X.; Hong, C. A sandwich-type electrochemical immunosensor for ultrasensitive detection of multiple tumor markers using an electrical signal difference strategy. *Talanta* **2020**, *219*, 121322. [CrossRef]
56. Zhang, X.; Li, Y.; Lv, H.; Feng, J.; Gao, Z.; Wang, P.; Dong, Y.; Liu, Q.; Zhao, Z. Sandwich-type electrochemical immunosensor based on Au@ Ag supported on functionalized phenolic resin microporous carbon spheres for ultrasensitive analysis of α-fetoprotein. *Biosens. Bioelectron.* **2018**, *106*, 142–148. [CrossRef] [PubMed]
57. Zhang, C.; Zhang, S.; Jia, Y.; Li, Y.; Wang, P.; Liu, Q.; Xu, Z.; Li, X.; Dong, Y. Sandwich-type electrochemical immunosensor for sensitive detection of CEA based on the enhanced effects of Ag NPs@ CS spaced Hemin/rGO. *Biosens. Bioelectron.* **2019**, *126*, 785–791. [CrossRef] [PubMed]
58. Yola, M.L.; Atar, N. Amperometric galectin-3 immunosensor-based gold nanoparticle-functionalized graphitic carbon nitride nanosheets and core–shell Ti-MOF@ COFs composites. *Nanoscale* **2020**, *12*, 19824–19832. [CrossRef]
59. Yan, Q.; Cao, L.; Dong, H.; Tan, Z.; Hu, Y.; Liu, Q.; Liu, H.; Zhao, P.; Chen, L.; Liu, Y. Label-free immunosensors based on a novel multi-amplification signal strategy of TiO2-NGO/Au@ Pd hetero-nanostructures. *Biosens. Bioelectron.* **2019**, *127*, 174–180. [CrossRef]
60. Martínez-Periñán, E.; Sánchez-Tirado, E.; González-Cortés, A.; Barderas, R.; Sánchez-Puelles, J.M.; Martínez-Santamaría, L.; Campuzano, S.; Yáñez-Sedeño, P.; Pingarrón, J. Amperometric determination of endoglin in human serum using disposable immunosensors constructed with poly (pyrrolepropionic) acid-modified electrodes. *Electrochim. Acta* **2018**, *292*, 887–894. [CrossRef]
61. Ehzari, H.; Samimi, M.; Safari, M.; Gholivand, M.B. Label-free electrochemical immunosensor for sensitive HER2 biomarker detection using the core-shell magnetic metal-organic frameworks. *J. Electroanal. Chem.* **2020**, *877*, 114722. [CrossRef]
62. Hou, Y.-H.; Wang, J.-J.; Jiang, Y.-Z.; Lv, C.; Xia, L.; Hong, S.-L.; Lin, M.; Lin, Y.; Zhang, Z.-L.; Pang, D.-W. A colorimetric and electrochemical immunosensor for point-of-care detection of enterovirus 71. *Biosens. Bioelectron.* **2018**, *99*, 186–192. [CrossRef]
63. Razzino, C.A.; Serafín, V.; Gamella, M.; Pedrero, M.; Montero-Calle, A.; Barderas, R.; Calero, M.; Lobo, A.O.; Yáñez-Sedeño, P.; Campuzano, S. An electrochemical immunosensor using gold nanoparticles-PAMAM-nanostructured screen-printed carbon electrodes for tau protein determination in plasma and brain tissues from Alzheimer patients. *Biosens. Bioelectron.* **2020**, *163*, 112238. [CrossRef]

64. Guan, J.-G.; Miao, Y.-Q.; Zhang, Q.-J. Impedimetric biosensors. *J. Biosci. Bioeng.* **2004**, *97*, 219–226. [CrossRef]
65. Bhand, S.; Bacher, G. Impedimetric sensors in environmental analysis: An overview. In *Environmental, Chemical and Medical Sensors*; Elsevier: Amsterdam, The Netherlands, 2018; pp. 67–85.
66. Manzoor, S.; Husain, S.; Somvanshi, A.; Fatema, M. Investigation of relaxation phenomenon in lanthanum orthoferrite extracted through complex impedance and electric modulus spectroscopy. *J. Appl. Phys.* **2020**, *128*, 064101. [CrossRef]
67. Prodromidis, M.I. Impedimetric immunosensors—A review. *Electrochim. Acta* **2010**, *55*, 4227–4233. [CrossRef]
68. Le, H.T.N.; Kim, J.; Park, J.; Cho, S. A review of electrical impedance characterization of cells for label-free and real-time assays. *BioChip J.* **2019**, *13*, 295–305.
69. Castizo-Olier, J.; Irurtia, A.; Jemni, M.; Carrasco-Marginet, M.; Fernandez-Garcia, R.; Rodriguez, F.A. Bioelectrical impedance vector analysis (BIVA) in sport and exercise: Systematic review and future perspectives. *PLoS ONE* **2018**, *13*, e0197957.
70. Grossi, M.; Riccò, B. Electrical impedance spectroscopy (EIS) for biological analysis and food characterization: A review. *J. Sens. Sens. Syst.* **2017**, *6*, 303–325. [CrossRef]
71. Zhang, Y.; Tang, Q.; Zhang, Y.; Wang, J.; Stimming, U.; Lee, A.A. Identifying degradation patterns of lithium ion batteries from impedance spectroscopy using machine learning. *Nat. Commun.* **2020**, *11*, 1706. [CrossRef]
72. Aydin, E.B.; Aydin, M.; Sezginturk, M.K. Advances in electrochemical immunosensors. *Adv. Clin. Chem.* **2019**, *92*, 1–57.
73. Han, E.; Li, X.; Zhang, Y.; Zhang, M.; Cai, J.; Zhang, X. Electrochemical immunosensor based on self-assembled gold nanorods for label-free and sensitive determination of *Staphylococcus aureus*. *Anal. Biochem.* **2020**, *611*, 113982. [CrossRef]
74. Malla, P.; Chen, G.-C.; Liao, H.-P.; Liu, C.-H.; Wu, W.-C. Label-free parathyroid hormone immunosensor using nanocomposite modified carbon electrode. *J. Electroanal. Chem.* **2021**, *880*, 114917. [CrossRef]
75. Nawaz, M.H.; Hayat, A.; Catanante, G.; Latif, U.; Marty, J.L. Development of a portable and disposable NS1 based electrochemical immunosensor for early diagnosis of dengue virus. *Anal. Chim. Acta* **2018**, *1026*, 1–7. [CrossRef]
76. Nessark, F.; Eissa, M.; Baraket, A.; Zine, N.; Nessark, B.; Zouaoui, A.; Bausells, J.; Errachid, A. Capacitance polypyrrole-based impedimetric immunosensor for interleukin-10 cytokine detection. *Electroanalysis* **2020**, *32*, 1795–1806. [CrossRef]
77. Sadighbayan, D.; Sadighbayan, K.; Tohid-Kia, M.R.; Khosroushahi, A.Y.; Hasanzadeh, M. Development of electrochemical biosensors for tumor marker determination towards cancer diagnosis: Recent progress. *TrAC Trends Anal. Chem.* **2019**, *118*, 73–88. [CrossRef]
78. Aydın, M.; Aydın, E.B.; Sezgintürk, M.K. A highly selective electrochemical immunosensor based on conductive carbon black and star PGMA polymer composite material for IL-8 biomarker detection in human serum and saliva. *Biosens. Bioelectron.* **2018**, *117*, 720–728. [CrossRef] [PubMed]
79. Aydın, E.B.; Aydın, M.; Sezgintürk, M.K. Electrochemical immunosensor based on chitosan/conductive carbon black composite modified disposable ITO electrode: An analytical platform for p53 detection. *Biosens. Bioelectron.* **2018**, *121*, 80–89. [CrossRef] [PubMed]
80. Aydın, M.; Aydın, E.B.; Sezgintürk, M.K. A disposable immunosensor using ITO based electrode modified by a star-shaped polymer for analysis of tumor suppressor protein p53 in human serum. *Biosens. Bioelectron.* **2018**, *107*, 1–9. [CrossRef]
81. Aydın, E.B.; Aydın, M.; Sezgintürk, M.K. Fabrication of electrochemical immunosensor based on acid-substituted poly (pyrrole) polymer modified disposable ITO electrode for sensitive detection of CCR4 cancer biomarker in human serum. *Talanta* **2021**, *222*, 121487. [CrossRef]
82. Aydın, E.B.; Aydın, M.; Sezgintürk, M.K. Highly selective and sensitive sandwich immunosensor platform modified with MUA-capped GNPs for detection of spike Receptor Binding Domain protein: A precious marker of COVID 19 infection. *Sens. Actuators B Chem.* **2021**, *345*, 130355. [CrossRef]
83. Simão, E.P.; Frías, I.A.; Andrade, C.A.; Oliveira, M.D. Nanostructured electrochemical immunosensor for detection of serological alkaline phosphatase. *Colloids Surf. B. Biointerfaces* **2018**, *171*, 413–418. [CrossRef]
84. Vasantham, S.; Alhans, R.; Singhal, C.; Nagabooshanam, S.; Nissar, S.; Basu, T.; Ray, S.C.; Wadhwa, S.; Narang, J.; Mathur, A. Paper based point of care immunosensor for the impedimetric detection of cardiac troponin I biomarker. *Biomed. Microdevices* **2020**, *22*, 6. [CrossRef]
85. Babamiri, B.; Bahari, D.; Salimi, A. Highly sensitive bioaffinity electrochemiluminescence sensors: Recent advances and future directions. *Biosens. Bioelectron.* **2019**, *142*, 111530. [CrossRef]
86. Li, L.; Chen, Y.; Zhu, J.J. Recent advances in electrochemiluminescence analysis. *Anal. Chem.* **2017**, *89*, 358–371. [CrossRef]
87. Richter, M.M. Electrochemiluminescence (ecl). *Chem. Rev.* **2004**, *104*, 3003–3036. [CrossRef] [PubMed]
88. Zhuo, Y.; Wang, H.J.; Lei, Y.M.; Zhang, P.; Liu, J.L.; Chai, Y.Q.; Yuan, R. Electrochemiluminescence biosensing based on different modes of switching signals. *Analyst* **2018**, *143*, 3230–3248. [CrossRef]
89. Bard, A.J. *Electrogenerated Chemiluminescence*; CRC Press: Boca Raton, FL, USA, 2004.
90. Hu, L.; Xu, G. Applications and trends in electrochemiluminescence. *Chem. Soc. Rev.* **2010**, *39*, 3275–3304. [CrossRef]
91. Fähnrich, K.A.; Pravda, M.; Guilbault, G.G. Recent applications of electrogenerated chemiluminescence in chemical analysis. *Talanta* **2001**, *54*, 531–559. [CrossRef]
92. Pyati, R.; Richter, M.M. ECL—Electrochemical luminescence. *Ann. Rep. Sect. C Phys. Chem.* **2007**, *103*, 12–78. [CrossRef]
93. Muzyka, K. Current trends in the development of the electrochemiluminescent immunosensors. *Biosens. Bioelectron.* **2014**, *54*, 393–407. [CrossRef]
94. Miao, W. Electrogenerated chemiluminescence and its biorelated applications. *Chem. Rev.* **2008**, *108*, 2506–2553. [CrossRef]

95. Marquette, C.A.; Blum, L.J. Electro-chemiluminescent biosensing. *Anal. Bioanal. Chem.* **2008**, *390*, 155–168. [CrossRef] [PubMed]
96. Forster, R.J.; Bertoncello, P.; Keyes, T.E. Electrogenerated chemiluminescence. *Annu. Rev. Anal. Chem.* **2009**, *2*, 359–385. [CrossRef]
97. Bronstein, I.; Olesen, C.E. Detection methods using chemiluminescence. In *Molecular Methods for Virus Detection*; Elsevier: Amsterdam, The Netherlands, 1995; pp. 147–174.
98. Shipkova, M.; Vogeser, M.; Ramos, P.A.; Verstraete, A.G.; Orth, M.; Schneider, C.; Wallemacq, P. Multi-center analytical evaluation of a novel automated tacrolimus immunoassay. *Clin. Biochem.* **2014**, *47*, 1069–1077. [CrossRef]
99. Liu, L.; Yang, A.; Luo, W.; Liu, H.; Liu, X.; Zhao, W. Ultrasensitive detection of cyclin D1 by a self-enhanced ECL immunosensor based on Bi 2 S 3 quantum dots. *Analyst* **2021**, *146*, 2057–2064. [CrossRef]
100. Du, D.; Shu, J.; Guo, M.; Haghighatbin, M.A.; Yang, D.; Bian, Z.; Cui, H. Potential-resolved differential electrochemiluminescence immunosensor for cardiac troponin I based on MOF-5-wrapped CdS quantum dot nanoluminophores. *Anal. Chem.* **2020**, *92*, 14113–14121. [CrossRef]
101. Wang, H.; Chai, Y.; Li, H.; Yuan, R. Sensitive electrochemiluminescent immunosensor for diabetic nephropathy analysis based on tris (bipyridine) ruthenium (II) derivative with binary intramolecular self-catalyzed property. *Biosens. Bioelectron.* **2018**, *100*, 35–40. [CrossRef]
102. Yang, H.; Wang, H.; Xiong, C.; Chai, Y.; Yuan, R. Highly sensitive electrochemiluminescence immunosensor based on ABEI/H_2O_2 system with PFO dots as enhancer for detection of kidney injury molecule-1. *Biosens. Bioelectron.* **2018**, *116*, 16–22. [CrossRef]
103. Zheng, X.; Mo, G.; He, Y.; Qin, D.; Jiang, X.; Mo, W.; Deng, B. An electrochemiluminescence immunosensor based on ZnSe@ ZnS QDs composite for CEA detection in human serum. *J. Electroanal. Chem.* **2019**, *844*, 132–141. [CrossRef]
104. Lian, X.; Feng, Z.; Tan, R.; Mi, X.; Tu, Y. Direct electrochemiluminescent immunosensing for an early indication of coronary heart disease using dual biomarkers. *Anal. Chim. Acta* **2020**, *1110*, 82–89. [CrossRef]
105. Qin, D.; Jiang, X.; Mo, G.; Feng, J.; Deng, B. Boron nitride quantum dots as electrochemiluminescence coreactants of rGO@ Au@ Ru–SiO_2 for label-free detection of AFP in human serum. *Electrochim. Acta* **2020**, *335*, 135621. [CrossRef]
106. Khan, M.S.; Ameer, H.; Chi, Y. Label-free and ultrasensitive electrochemiluminescent immunosensor based on novel luminophores of $Ce_2Sn_2O_7$ nanocubes. *Anal. Chem.* **2021**, *93*, 3618–3625. [CrossRef] [PubMed]
107. Babamiri, B.; Hallaj, R.; Salimi, A. Ultrasensitive electrochemiluminescence immunosensor for determination of hepatitis B virus surface antigen using CdTe@ CdS-PAMAM dendrimer as luminescent labels and Fe_3O_4 nanoparticles as magnetic beads. *Sens. Actuators B Chem.* **2018**, *254*, 551–560. [CrossRef]
108. Fang, D.; Zhang, S.; Dai, H.; Lin, Y. An ultrasensitive ratiometric electrochemiluminescence immunosensor combining photothermal amplification for ovarian cancer marker detection. *Biosens. Bioelectron.* **2019**, *146*, 111768. [CrossRef] [PubMed]
109. Ding, J.; Jiang, W.; Zhou, Y.; Yin, H.; Ai, S. Electrochemiluminescence immunosensor for 5-hydroxymethylcytosine detection based on PAMAM-nanosilver-nitrogen doped graphene nanocomposite. *J. Electroanal. Chem.* **2020**, *877*, 114646. [CrossRef]
110. Yang, C.; Guo, Q.; Lu, Y.; Zhang, B.; Nie, G. Ultrasensitive "signal-on" electrochemiluminescence immunosensor for prostate-specific antigen detection based on novel nanoprobe and poly (indole-6-carboxylic acid)/flower-like Au nanocomposite. *Sens. Actuators B Chem.* **2020**, *303*, 127246. [CrossRef]

 biosensors

Article

High-Aspect-Ratio Microfluidic Channel with Parallelogram Cross-Section for Monodisperse Droplet Generation

Hyeonyeong Ji [1], Jaehun Lee [2], Jaewon Park [3], Jungwoo Kim [1], Hyun Soo Kim [4,*] and Younghak Cho [1,*]

1 Department of Mechanical System Design Engineering, Seoul National University of Science & Technology, Seoul 01811, Korea; jhy8718@naver.com (H.J.); kimjw@seoultech.ac.kr (J.K.)
2 Daegu Research Center for Medical Devices and Rehabilitation, Korea Institute of Machinery and Materials, Daegu 42994, Korea; ljh30226@naver.com
3 School of Microelectronics, Southern University of Science and Technology, Shenzhen 518055, China; jwpark@sustech.edu.cn
4 Department of Electronic Engineering, Kwangwoon University, Seoul 01897, Korea
* Correspondence: hyunsookim@kw.ac.kr (H.S.K.); yhcho@seoultech.ac.kr (Y.C.)

Abstract: Droplet-based microfluidics has been widely used as a potent high-throughput platform due to various advantages, such as a small volume of reagent consumption, massive production of droplets, fast reaction time, and independent control of each droplet. Therefore, droplet microfluidic systems demand the reliable generation of droplets with precise and effective control over their size and distribution, which is critically important for various applications in the fields of chemical analysis, material synthesis, lab-on-a-chip, cell research, diagnostic test, and so on. In this study, we propose a microfluidic device with a high-aspect-ratio (HAR) channel, which has a parallelogram cross-section, for generating monodisperse droplets. The HAR channel was fabricated using simple and cheap MEMS processes, such as photolithography, anisotropic wet etching, and PDMS molding, without expensive equipment. In addition, the parallelogram cross-section channel structure, regarded as a difficult shape to implement in previous fabrication methods, was easily formed by the self-alignment between the silicon channel and the PDMS mold, both of which were created from a single crystal silicon through an anisotropic etching process. We investigated the effects of the cross-sectional shape (parallelogram vs. rectangle) and height-to-width ratio of microfluidic channels on the size and uniformity of generated droplets. Using the developed HAR channel with the parallelogram cross-section, we successfully obtained smaller monodisperse droplets for a wider range of flow rates, compared with a previously reported HAR channel with a rectangular cross-section.

Keywords: high-aspect-ratio microfluidic channel; parallelogram cross-section; monodisperse droplet; droplet generation

Citation: Ji, H.; Lee, J.; Park, J.; Kim, J.; Kim, H.S.; Cho, Y. High-Aspect-Ratio Microfluidic Channel with Parallelogram Cross-Section for Monodisperse Droplet Generation. *Biosensors* **2022**, *12*, 118. https://doi.org/10.3390/bios12020118

Received: 5 January 2022
Accepted: 8 February 2022
Published: 14 February 2022

1. Introduction

Droplets have been widely used as biochemical reactors for chemical and biological analysis and templates for polymer microbeads due to their unique features, such as high-throughput, minimal reagent consumption, massive production of droplets, rapid response time, and independent control of each droplet [1–5]. Therefore, droplet microfluidic systems require the reliable generation of droplets with precise and effective control over their size and distribution, which is critically important for various applications in the fields of chemical analysis, material synthesis, lab-on-a-chip, cell research, diagnostic test, and so on.

Compared with conventional droplet production processes, such as mechanical agitation [6] and membrane emulsification [7], methods based on microfluidic channels have several advantages of easy and robust generation of droplets, excellent handling, and monodisperse droplet size ranging from nanometer to micrometer scale [8]. There are various types of microfluidic channel configurations according to the droplet breakup mechanism: T-junction [9,10] and flow-focusing [11] configuration based on shear stress for droplet breakup and step emulsification [12,13], gradient of confinement [14], and HAR (high-aspect-ratio) confinement [15] configuration based on interfacial tension for droplet pinch-off. Using the T-junction structure, fluids with two immiscible phases generate droplets by shear force, whose sizes can be adjusted by changing the relative flow rates of these fluids. In order to generate a large amount of droplets with a uniform size, however, it is necessary to precisely control the flow rate, and thus, microfluidic devices having a precise flow control capability are required. On the other hand, in the step emulsification configuration, the interfacial tension between two immiscible phases by a Laplace pressure difference drives droplet pinch-off when passing through the stepped channel. This droplet breakup mechanism can produce uniform-sized droplets regardless of the interference of flow or pressure fluctuation, enabling massive droplet generation by parallelizing the multiple-step emulsification configurations on a single device. However, one of the drawbacks is that this configuration requires complicated multistep photolithography processes to fabricate stepped channels. Recently, Yao et al. reported a HAR microfluidic channel device that has a similar droplet generation mechanism to the step emulsification device [15]. As the fluid exits the HAR channel into the chamber, the strong constraints of the interface are released, and droplets are generated by the surface tension. In this case, the size of the droplet is determined by the channel width. That is, uniform droplets can be generated without being affected by the flow velocity when the aspect ratio (AR, the ratio of width to height) exceeds 3.5. In order to fabricate such HAR microfluidic channels, however, it requires the use of expensive deep reactive ion etching (RIE) equipment or a UV-based LIGA process [16].

In this study, we present a microfluidic device with a high-aspect-ratio (HAR) channel, which has a parallelogram cross-section, for generating monodisperse droplets. The developed HAR channel was fabricated using simple and cheap MEMS processes, such as photolithography, anisotropic KOH (potassium hydroxide) wet etching, and polydimethylsiloxane (PDMS) molding. Compared with the previously used HAR channel fabrication methods, the developed HAR channel was successfully fabricated without expensive equipment, and more importantly, the parallelogram cross-section channel structure, regarded as a very difficult structure to implement previously, was successfully integrated into a microfluidic device, particularly droplet generation configuration. In addition, compared with the above-mentioned HAR channel with a rectangular cross-section [15,16], our presented HAR channel with a parallelogram cross-section can generate more stable and smaller droplets for a wider range of flow rates. Here, we investigate the effects of the cross-sectional shape (parallelogram vs. rectangle) and height-to-width ratio (AR) of a microfluidic channel on the size and uniformity of generated droplets. To the best of our knowledge, the effect of the cross-sectional shape on the droplet generation has not been examined yet, and the developed microfluidic device is the first platform that employs the parallelogram cross-section channel in droplet production.

2. Materials and Methods

2.1. Design of a High-Aspect-Ratio (HAR) Microfluidic Channel with a Parallelogram Cross-Section

The developed microfluidic device is composed of two inlets for sample solution and carrier oil solution, an interconnecting channel for generating droplets at the interface of the oil inlet (T-junction droplet generation region), and an outlet for collecting the generated droplets (Figure 1). The middle interconnecting channel is designed to have a HAR geometry of a parallelogram cross-section, where its base width (W), height (H), and hypotenuse (HT) are 13.5, 51.0, and 62.4 µm, respectively, displaying aspect ratios (ARs) of 3.8 (H/W) and 4.6 (HT/W). The angle (θ) between the base width and the side length always forms 54.7° since this is the angle between (111) and (100) of single crystal Si. When the HAR channel (AR > 3.5) is utilized to provide the sample solution in the T-junction configuration, the droplet breakup mechanism is mainly dominated by the interfacial tension, and the effect of the shear stress on the droplet formation can be neglected [12,13]. The developed microfluidic device employing the HAR channel with a parallelogram cross-section enables the sample solution to self-break up into monodisperse droplets without being affected by fluid properties and flow rates.

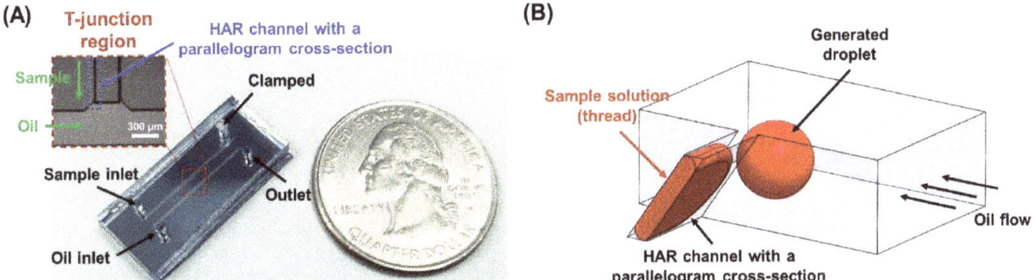

Figure 1. (**A**) Photographs of the developed microfluidic device comprising a HAR channel with a parallelogram cross-section. An inset shows the microscopic image of a T-junction droplet generation region in the microfluidic device where sample solution flows through the HAR channel with a parallelogram cross-section. (**B**) A schematic illustration of the HAR channel with a parallelogram cross-section in the T-junction region, which induces self-breakup droplet generation.

2.2. Fabrication Process

Previously, we reported the fabrication process of a channel with a parallelogram cross-section [17,18]. Photolithography, anisotropic KOH wet etching, plasma bonding, and self-alignment between PDMS and Si were sequentially performed to fabricate channels having parallelogram and rectangular cross-sections, as shown in Figure 2. First, a SiO_2 thin film layer of 1000 Å thickness was deposited on a (100) single crystal Si wafer using low-pressure chemical vapor deposition (LPCVD) and patterned by photolithography and RIE (reactive ion etch) (photomask designs are illustrated in Figure S1). Then, the Si wafer was anisotropically etched with KOH solution at 70 °C (etch rate: 0.66 μm/min). The Si channels and the masters for PDMS molds for parallelogram and rectangular cross-sections had the same etching depth, so it was possible to fabricate the Si channel and master for the PDMS mold in one silicon wafer. The width of the channels ($W = W_1 - W_2$), which was determined by the photomask design for Si channel width (W_1) and master width (W_2) for PDMS molds, was fixed to 13.5 μm, and their heights (H) were controlled by KOH anisotropic wet etching time to fabricate the channels with various ARs. A PDMS mold was replicated from the Si master, and then was self-aligned and bonded with the Si channel by oxygen plasma treatment. A small amount of methanol (or DI water) was sprayed between the Si channel and the PDMS mold to facilitate self-alignment, where the PDMS mold could be mechanically aligned to one side wall of the Si channel by hand. Finally, the methanol was evaporated on a hot plate to complete the formation of the channel, which was composed of PDMS and silicon. Since the Si channel and the Si master for the PDMS mold were fabricated on the same wafer, both had the same crystal plane and etching depth (Figure 2). In other words, the geometrical similarity (including depth) between them enabled the easy alignment. The assembled microfluidic device is illustrated in Figure 1A.

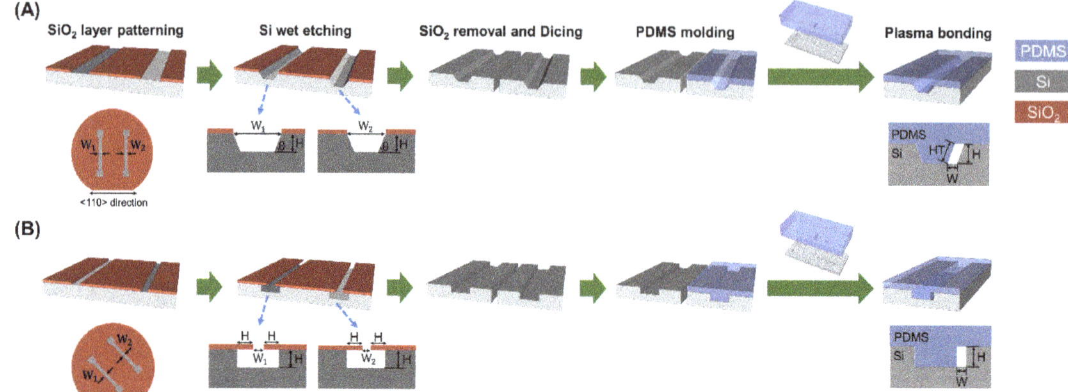

Figure 2. The fabrication process of the microfluidic devices consisting of a HAR channel (**A**) with a parallelogram cross-section and (**B**) with a rectangular cross-section. The Si channel and the Si master for the PDMS mold are fabricated on a single Si wafer, allowing for both the Si channel and the PDMS mold to have the same height. The channel width (W) in both parallelogram and rectangular cross-section designs was determined by an initial photomask design, resulting in $W = W_1 - W_2$. The channel height (H) can be easily controlled by adjusting the etching time.

2.3. Droplet Generation and Analysis

Channels in all fabricated devices were coated with Aquapel® (PGW LLC, Cranberry Township, PA, USA) to make their surfaces hydrophobic, followed by rinsing with nitrogen gas and carrier oil (Novec™ 7500, 3M™, St. Paul, MN, USA). To flow the sample solution (deionized water) into the HAR channel, one of two inlets was clamped. Oil solution was mixed with a surfactant (008-FluoroSurfactant, RAN Biotechnologies, Beverly, MA, USA) at a 1% (w/w) ratio for stable droplet generation and storage. Both sample and oil solutions were injected using syringe pumps (Fusion 200, Chemyx Inc., Stafford, TX, USA).

To compare the effect of cross-sectional shapes (parallelogram or rectangle) and ARs on the droplet generation, microfluidic devices with 3 different interconnecting channel designs (P–W13.5–H29.5, P–W13.5–H51.0, and R–W13.5–H51.0; having the same channel width (W) but different shapes (P: parallelogram, R: rectangle) and heights (H), Figure 3) were characterized. The flow rate of the carrier oil solution was fixed at 400 µL/h, and droplet generation under various flow rates of the sample solution (10, 30, 50, 100, and 200 µL/h) was analyzed. For further precise comparison between P–W13.5–H51.0 and R–W13.5–H51.0, even lower flow rates of the sample solution (1, 2, and 5 µL/h) were tested. The hydraulic diameters (D_h) of 3 different interconnecting channel designs were calculated from the definition of Reynolds number (Re) as below:

$$Re = \frac{\rho V D_h}{\mu} = \frac{\rho \frac{Q}{A}\frac{4A}{P}}{\mu} = \frac{4\rho}{\mu}\frac{Q}{P} \quad \left(V = \frac{Q}{A},\ \ D_h = \frac{4A}{P} \right)$$

where ρ is the fluid density, V is the characteristic velocity, D_h is the hydraulic diameter, μ is the mean viscosity, A is the cross-sectional area of the flow, P is the wetted perimeter of the cross-section, and Q is the flow rate.

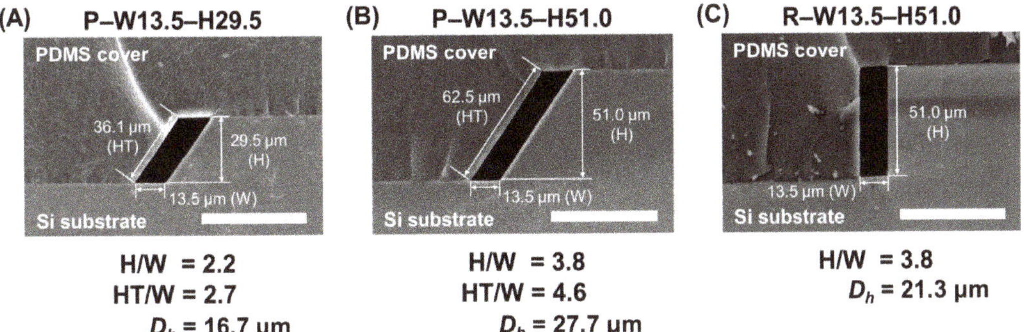

Figure 3. The SEM images showing the cross-section of the fabricated HAR channels. (**A**) Cross-section of a parallelogram channel with AR lower than 3.5 (AR = 2.2, P–W13.5–H29.5). (**B**) Cross-section of a parallelogram channel with AR larger than 3.5 (AR = 3.8, P–W13.5–H51.0). (**C**) Cross-section of a rectangular channel with AR larger than 3.5 (AR = 3.8, R–W13.5–H51.0) (scale bar: 100 µm).

The developed HAR channel has a parallelogram cross-section structure, which is not symmetric, and depending on the flow direction of oil solution, the sample solution would experience different flow profiles at the interface (i.e., T-junction region). For example, when the oil solution flows towards the outlet, this carrier solution interacts with the sample solution with an angle of 54.7°, while the oil solution meets the sample solution with an angle of 125.3° when flowing towards the inlet. In order to examine the effect of the different flow profiles at the interface, the droplet production under different oil flow directions was compared (oil flow rate: 400 µL/h). In addition, flow rates of 200, 400, 600, and 800 µL/h in the oil solution were tested to investigate their influence on the produced droplet size. The flow rate of the sample solution was maintained at 50 µL/h. For this characterization, a HAR channel with a parallelogram cross-section (P–W11.8–H51.0; W: 11.8 µm, H: 51.0 µm, HT: 62.4 µm, H/W: 4.3, HT/W: 5.3) was utilized.

Droplet generation was monitored using an upright microscope (BXFM-F, Olympus, Tokyo, Japan) equipped with a high-speed camera (MIRO EX4-4096MC, Phantom, Wayne, NJ, USA). The generated droplets were collected through the outlet, and their sizes were measured by bright-field microscopy and NIH ImageJ software. Each datum shown in the manuscript was measured from at least 30 droplets.

3. Results and Discussion

3.1. High-Aspect-Ratio Microfluidic Channel with a Parallelogram Cross-Section

Figure 3 shows the scanning electron microscope (SEM) images of the fabricated channels having parallelogram and rectangular cross-sections. These cross-sectional images clearly demonstrate that the Si channel and the PDMS mold (replicated from the Si master) were perfectly aligned and bonded because of their geometrical similarity. The width of both fabricated channels was found to be 13.5 µm, which was successfully determined by the width of the initial photomask design. The height of the fabricated channels was easily controlled by adjusting the anisotropic KOH etching time. For example, channels in Figure 3A,B, both have a parallelogram cross-section and the same channel width of 13.5 µm, but exhibit different channel heights, 29.5 and 51.0 µm (H/W = 2.2 and 3.8), where a deeper channel height was achieved through the longer wet etching process. Figure 3C illustrates the cross-section of the fabricated rectangular interconnecting channel, which has the same width and height, resulting in the same AR (H/W = 3.8) compared with Figure 3B. Both channels have the same cross-sectional areas and Reynolds number for a given flow rate, but the hydraulic diameter is different. It means that the fluid would experience the different force when it flows out from the channel.

The HAR channel with a parallelogram cross-section was successfully fabricated using simple and cheap MEMS processes without requiring any expensive equipment. In addition, the parallelogram channel shape, which was challenging to implement in conventional fabrication methods, was built and integrated into the microfluidic device. To the best of our knowledge, the developed microfluidic device is the first platform that utilizes the parallelogram cross-section channel for droplet generation.

3.2. Characterization of Droplet Generation with Different Channel Geometries

Time-lapse images showing the droplet generation process at the water–oil interface are illustrated in Figure 4. When the sample solution (DI water) was introduced in the HAR channel with a parallelogram cross-section, it was pressurized and formed a thin thread. As the sample solution reached close to the T-junction region, the thin thread pinched off into a droplet. Once the formed droplet was released to an oil-flowing channel, the elongated thread retracted, resumed the initial state, and repeated these steps, resulting in monodisperse droplets being produced periodically (Video S1).

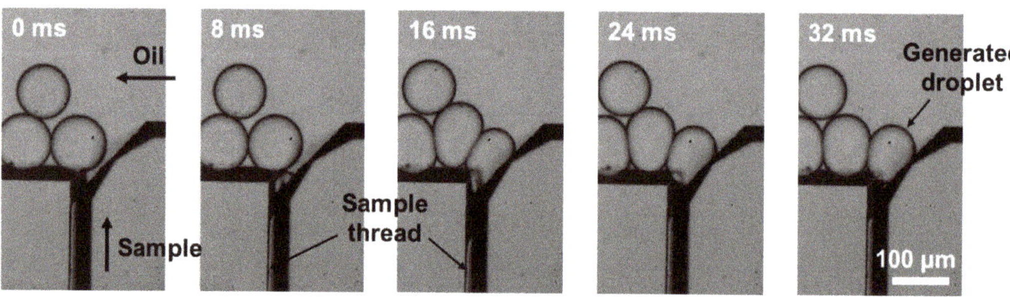

Figure 4. Time-lapse images displaying the droplet generation process at the T-junction region.

The effect of the channel geometries, such as cross-sectional shapes (parallelogram or rectangle) and ARs (larger or smaller than 3.5), on droplet formation was investigated. The diameters of the droplets created from all the different geometries and flow rate conditions (oil: 400 μL/h, sample: 1, 2, 5, 10, 30, 50, 100, and 200 μL/h) were measured and compared, where all measurement results are summarized in Table 1. It can be seen in Figure 5A,B that the modality of the droplet formation is different in accordance with the channel geometries. For example, when the channel structure with an AR value of less than 3.5 (P–W13.5–H29.5, AR = 2.2) was used, the droplet breakup profile followed a conventional T-junction droplet generation principle, which is governed by the viscous shearing forces between oil and sample solutions. In this case, the size of the produced droplet was strongly dependent on flow variations, where larger-diameter droplets were formed as the faster flow of the sample solution was applied (Video S2). In the P–W13.5–H29.5 channel design, the average diameter of the droplets was 46.2 μm under the sample flow rate of 10 μL/h, and then their sizes began to increase as the sample flow rates became larger, where droplets with an average diameter of 146.3 μm were created under a sample flow rate of 200 μL/h.

On the other hand, when the HAR channel structure (AR > 3.5) was employed, the interfacial tension became the dominant force for droplet generation, enabling monodisperse droplets to be formed regardless of flow fluctuations. In the P–W13.5–H51.0 channel design (AR = 3.8), the average sizes of the droplets were kept almost consistent, where less than 1.5% size variation was observed between sample flow rates ranging from 10 to 100 μL/h (Table 1, Figure 5B, and Video S3). In the R–W13.5–H51.0 channel design (AR = 3.8), the average diameters of the droplets created under the sample flow rate conditions ranging from 10 to 50 μL/h were also almost uniform (Table 1, Figure 5B, and Video S4), although this rectangular cross-section design had a smaller flow rate interval for inducing the monodisperse droplet generation than the parallelogram cross-section design (P–W13.5–H51.0).

Table 1. Average diameter of droplets generated from different channel geometries and flow rate conditions.

		Sample Flow Rate (μL/h)							
		1	2	5	10	30	50	100	200
Channel geometry	**P–W13.5–H29.5 (parallelogram, AR < 3.5, D_h = 16.7 μm)**				46.2 ± 1.4	66.3 ± 2.3	72.0 ± 2.9	118.5 ± 5.6	146.3 ± 24.7
	P–W13.5–H51.0 (parallelogram, AR > 3.5, D_h = 27.7 μm)	45.3 ± 0.7	45.3 ± 0.7	45.6 ± 1.3	46.3 ± 1.0	46.9 ± 0.9	46.8 ± 0.9	46.9 ± 2.0	151.9 ± 60.9
	R–W13.5–H51.0 (rectangle, AR > 3.5, D_h = 21.3 μm)	75.8 ± 1.6	77.8 ± 1.6	81.2 ± 1.5	90.0 ± 4.6	90.3 ± 3.2	91.4 ± 2.1	111.7 ± 5.2	149.2 ± 10.5

(unit: μm).

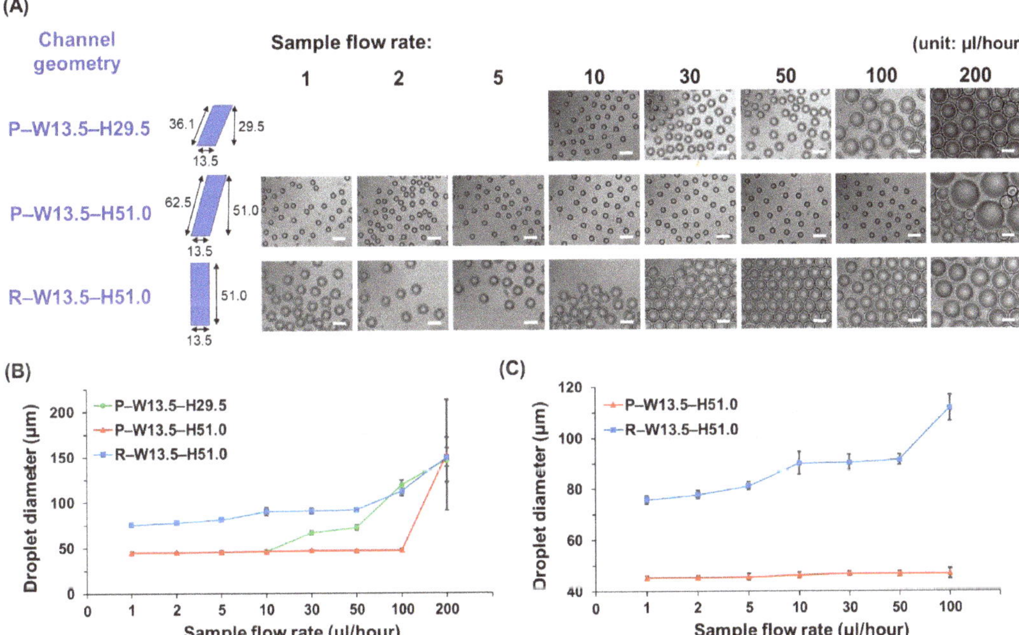

Figure 5. The effect of channel geometries (cross-section shape and AR) on droplet generation. (**A**) Micrographs showing the droplets produced using different channel geometries under various flow rate conditions of the sample solution. The flow rate of the oil solution was fixed at 400 μL/h (scale bar: 100 μm). (**B**,**C**) Analysis of the average droplet sizes by different geometries and sample flow rate conditions.

To further analyze the effect of cross-section shapes, even lower flow rates (1, 2, and 5 μL/h) of sample solutions were applied, and the droplet production profiles were compared between the P–W13.5–H51.0 and R–W13.5–H51.0 channel designs. As shown in Figure 5C, the parallelogram cross-section design (P–W13.5–H51.0) was able to produce almost uniform-sized droplets irrespective of the flow rate changes even in this lower range. However, the average droplet diameters created from the rectangular cross-section design (R–W13.5–H51.0) showed the dependency on the flow variations (e.g., 75.8 μm at 1 μL/h and 81.2 μm at 5 μL/h), indicating that the viscous shearing force was the dominant force in the lower range, and uniform-sized droplets could not be implemented. In addition, the hydraulic diameters (D_h) of three different channel designs were calculated and compared (Table 1 and Figure 3), where the channel design with a larger hydraulic diameter (P–W13.5–H51.0, D_h = 27.7 μm) showed the generation of monodisperse droplets within a wider range of flow rates against other designs (P–W13.5–H29.5, D_h = 16.0 μm, R–W13.5–H51.0, D_h = 21.3 μm). From the above characterization results of droplet generation among three different geometries, our developed HAR channel with a parallelogram cross-section clearly demonstrated its outperforming capability, which could generate uniform-sized droplets independent of flow variation within a wider range (flow rate range for uniform-sized droplet generation: the developed HAR parallelogram channel = 1 ~ 100 μL/h vs. the HAR rectangular channel = 10 ~ 50 μL/h).

Another interesting feature of the parallelogram cross-section design is that the produced droplets have smaller sizes compared with the rectangular cross-section design. In the P–W13.5–H51.0 and R–W13.5–H51.0 channels, both designs have the same channel width and cross-sectional area, so they would have to generate similar sizes of droplets under the same flow conditions. However, when comparing the droplet sizes created under the same flow conditions, the parallelogram cross-section design (P–W13.5–H51.0) exhibits smaller sizes all the time. This size difference would be mainly due to different cross-sectional shapes, particularly the hypotenuse of the parallelogram. When the sample solution is filled inside the P–W13.5–H51.0 channel, the channel surface that interacts with the solution thread to produce droplets is the hypotenuse of the parallelogram, not its height. Thus, although the parallelogram cross-section channel has the same channel height (same AR) and cross-sectional area compared with the rectangular cross-section channel, this design can provide a larger surface interaction (the length of the hypotenuse = the channel height/sin (54.7°) = 1.23 × the channel height = 62.5 μm), which would result in a larger interfacial tension. The larger interfacial tension might induce more frequent droplet breakups along the parallelogram cross-section channel, and this would be the main reason for smaller droplets being produced with higher generation frequencies (Videos S3 and S4). In addition, the larger interfacial tension deriving from the interaction with a longer hypotenuse would be the main contributor for a wider range of flow variation that can create the uniform-sized droplets in the parallelogram cross-section channel. Numerical simulations based on the volume of fluid were also conducted to clearly understand the effect of the channel cross-section geometries on the droplet generation, where the difference in the droplet breakup mechanism as well as the sample flowing position among three different designs was observed (Supplementary Information, Figures S2 and S3).

3.3. Effect of the Carrier Oil on Droplet Generation

As the developed HAR channel has the parallelogram structure, the channel cross-section is tilted at an angle of 54.7° to the bottom surface. This asymmetric channel structure can cause different flow profiles at a sample (water)–oil interface when the oil flow directions change in the T-junction. To confirm whether the droplet generation was affected by the oil flow directions, the flow rates of the sample and oil solutions were fixed at 50 and 400 μL/h, respectively, and the droplet formation was monitored only by varying the oil flow direction (Figure 6A). As can be seen in Figure 6B and Video S5, no significant difference in droplet generation (droplet diameter: 41.0 ± 0.8 μm vs. 41.2 ± 0.9 μm, corresponding to oil flow direction towards outlet (forward flow) vs. oil inlet (reverse flow)) was observed from different oil flow directions (i.e., different flow profiles). In addition, the effect of oil flow speed on the droplet production was tested (Figure 6C and Video S6), where no significant difference was found from various oil flow rates ranging from 200 to 800 μL/h (droplet diameter: 41.1 ± 0.8, 41.0 ± 0.8, 40.8 ± 0.7, and 41.4 ± 0.8 μm at oil flow rates of 200, 400, 600, and 800 μL/h, respectively). These results from different oil flow conditions verify that the droplet generation mechanism in the developed HAR parallelogram channel is mainly dominated by the interfacial tension, which induces self-breakup of droplets and is not affected by flow variation.

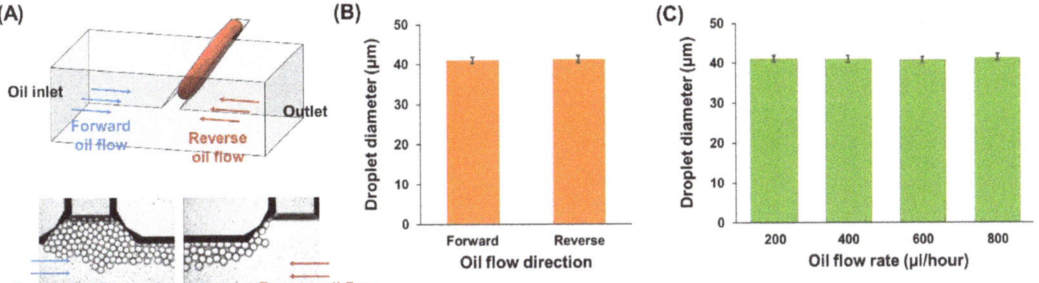

Figure 6. The effect of the carrier oil flow on droplet generation. (**A**) Illustration and microscopic images showing the droplet generation under two different flow directions of the oil solution (forward: towards a device outlet; reverse: towards an oil inlet) (scale bar: 100 μm). (**B**) Comparison of the average droplet sizes produced under two different oil flow conditions (sample flow rate: 50 μL/h; oil flow rate: 400 μL/h). (**C**) Analysis of the average droplet sizes formed under various oil flow rates (200, 400, 600, and 800 μL/h, sample flow rate was fixed at 50 μL/h).

One thing to note here is that the droplet size (diameter: 41.0 μm) generated from the HAR parallelogram channel having a 11.8 μm width is smaller than the previously used channel design (P–W13.5–H51.0) comprising a 13.5 μm width (droplet diameter: 46.8 μm). Since both channel designs had the same height and hypotenuse, the different droplet sizes were mainly derived from different lengths of the channel widths. This result clearly demonstrates the capability of the developed HAR parallelogram channels to control the droplet sizes by simply adjusting the channel width.

4. Conclusions

In this paper, we developed the microfluidic device comprising the high-aspect-ratio (HAR) channel with a parallelogram cross-section for generating monodisperse droplets. The developed device was fabricated using simple and cheap MEMS processes, where the HAR parallelogram channel geometry, very difficult to implement previously, was successfully created and integrated into the microfluidic device without requiring high-cost equipment and processes. The droplet generation was characterized using the developed channel design, and the effect of the channel geometries including the aspect-ratio and cross-sectional shape was investigated. In addition, the performance of the developed parallelogram channel was compared with that of a previously developed rectangular channel. The results successfully demonstrated the outperforming capability of the developed HAR parallelogram channel, where uniform-sized droplet generation was confirmed in a wider range of flow variation. To the best of our knowledge, the developed microfluidic device is the first platform that utilizes a parallelogram cross-section channel for generating droplets.

Supplementary Materials: The following supporting information can be downloaded at: https://www.mdpi.com/article/10.3390/bios12020118/s1, Figure S1. Photomask designs of microfluidic devices; Supplementary Information. Numerical simulations of the droplet generation in channels with parallelogram and rectangular cross-section; Figure S2. Contours of the void fraction of the dispersed phase in the cross plane; Figure S3. Contours of the void fraction of the dispersed phase in the top plane of the interconnecting channel; Video S1. Droplet generation at the water–oil interface (T-junction region) displaying the self-breakup into a monodisperse droplet; Video S2. Droplet generation in the P–W13.5–H29.5 channel (parallelogram, AR < 3.5) under various flow rate conditions; Video S3. Droplet generation in the P–W13.5–H51.0 channel (parallelogram, AR > 3.5) under various flow rate conditions; Video S4. Droplet generation in the R–W13.5–H51.0 channel (rectangle, AR > 3.5) under various flow rate conditions; Video S5. The effect of oil flow direction on the droplet generation in the HAR channel with a parallelogram cross-section (AR > 3.5); Video S6. The effect of oil flow rates on the droplet generation in the HAR channel with a parallelogram cross-section (AR > 3.5).

Author Contributions: J.P., H.S.K. and Y.C. conceived the design; H.J., J.L., J.K. and H.S.K. performed the experiments; H.J., J.P., J.K., H.S.K. and Y.C. analyzed the data; H.J., H.S.K. and Y.C. wrote the original draft, and all authors reviewed and edited the manuscript. H.S.K. and Y.C. supervised the overall work. All authors have read and agreed to the published version of the manuscript.

Funding: This work was supported by the Korea Institute of Radiological and Medical Sciences (50538-2020) funded by the Korean Ministry of Science and ICT. This research was also supported by the Korea Environmental Industry and Technology Institute (KEITI) through a project to develop new eco-friendly materials and a processing technology derived from wildlife, funded by the Korean Ministry of Environment (MOE) (2021003240004), and the National Research Foundation of Korea (NRF) grant funded by the Korean government (MSIT) (NRF-2021R1F1A1063455).

Institutional Review Board Statement: Not applicable.

Informed Consent Statement: Not applicable.

Conflicts of Interest: The authors declare no conflict of interest.

References

1. Sohrabi, S.; Kassir, N.; Moraveji, M.K. Droplet microfluidics: Fundamentals and its advanced applications. *RSC Adv.* **2020**, *10*, 27560–27574. [CrossRef]
2. Liu, Z.M.; Yang, Y.; Du, Y.; Pang, Y. Advances in droplet-based microfluidic technology and its applications. *Chin. J. Anal. Chem.* **2017**, *45*, 282–296. [CrossRef]
3. Chou, W.L.; Lee, P.Y.; Yang, C.L.; Huang, W.Y.; Lin, Y.S. Recent advances in applications of droplet microfluidics. *Micromachines* **2015**, *6*, 1249–1271. [CrossRef]
4. Kim, H.S.; Hsu, S.C.; Han, S.I.; Thapa, H.R.; Guzman, A.R.; Browne, D.R.; Tatli, M.; Devarenne, T.P.; Stern, D.B.; Han, A. High-throughput droplet microfluidics screening platform for selecting fast-growing and high lipid-producing microalgae from a mutant library. *Plant Direct* **2017**, *1*, e00011. [CrossRef] [PubMed]
5. Kim, H.S.; Waqued, S.C.; Nodurft, D.T.; Devarenne, T.P.; Yakovlev, V.V.; Han, A. Raman spectroscopy compatible PDMS droplet microfluidic culture and analysis platform towards on-chip lipidomics. *Analyst* **2017**, *142*, 1054–1060. [CrossRef] [PubMed]
6. Xu, Q.; Nakajima, M.; Ichikawa, S.; Nakamura, N.; Shiina, T. A comparative study of microbubble generation by mechanical agitation and sonication. *Innov. Food Sci. Emerg. Technol.* **2008**, *9*, 489–494. [CrossRef]
7. Joscelyne, S.M.; Trägårdh, G. Membrane emulsification—A literature review. *J. Membr. Sci.* **2000**, *169*, 107–117. [CrossRef]
8. Zhu, P.; Wang, L. Passive and active droplet generation with microfluidics: A review. *Lab Chip.* **2017**, *17*, 34–75. [CrossRef]
9. Liu, H.; Zhang, Y. Droplet formation in a T-shaped microfluidic junction. *J. Appl. Phys.* **2009**, *106*, 034906. [CrossRef]
10. Yao, J.; Lin, F.; Kim, H.S.; Park, J. The effect of oil viscosity on droplet generation rate and droplet size in a T-junction microfluidic droplet generator. *Micromachines* **2019**, *10*, 808. [CrossRef] [PubMed]
11. Yobas, L.; Martens, S.; Ong, W.L.; Ranganathan, N. High-performance flow-focusing geometry for spontaneous generation of monodispersed droplets. *Lab Chip* **2006**, *6*, 1073–1079. [CrossRef] [PubMed]
12. Shin, D.C.; Morimoto, Y.; Sawayama, J.; Miura, S.; Takeuchi, S. Centrifuge-based step emulsification device for simple and fast generation of monodisperse picoliter droplets. *Sens. Actuators B Chem.* **2019**, *301*, 127164. [CrossRef]
13. Shi, Z.; Lai, X.; Sun, C.; Zhang, X.; Zhang, L.; Pu, Z.; Wang, R.; Yu, H.; Li, D. Step emulsification in microfluidic droplet generation: Mechanisms and structures. *Chem. Commun.* **2020**, *56*, 9056–9066. [CrossRef] [PubMed]
14. Dangla, R.; Kayi, S.C.; Baroud, C.N. Droplet microfluidics driven by gradients of confinement. *Proc. Natl. Acad. Sci. USA* **2013**, *110*, 853–858. [CrossRef] [PubMed]
15. Xu, X.; Yuan, H.; Song, R.; Yu, M.; Chung, H.Y.; Hou, Y.; Shang, Y.; Zhou, H.; Yao, S. High aspect ratio induced spontaneous generation of monodisperse picolitre droplets for digital PCR. *Biomicrofluidics* **2018**, *12*, 014103. [CrossRef] [PubMed]
16. Chung, C.H.Y.; Cui, B.; Song, R.; Liu, X.; Xu, X.; Yao, S. Scalable production of monodisperse functional microspheres by multilayer parallelization of high aspect ratio microfluidic channels. *Micromachines* **2019**, *10*, 592. [CrossRef] [PubMed]
17. Kwon, J.Y.; Lee, D.K.; Cho, Y.H. Fabrication of microchannel with parallelogram cross-section using Si anisotropic wet etching and self-alignment. *J. Korean Soc. Precis. Eng.* **2019**, *36*, 287–291. [CrossRef]
18. Lee, D.K.; Kwon, J.Y.; Cho, Y.H. Fabrication of microfluidic channels with various cross-sectional shapes using anisotropic etching of Si and self-alignment. *Appl. Phys. A* **2019**, *125*, 291. [CrossRef]

 biosensors

Article

Urinary Volatiles and Chemical Characterisation for the Non-Invasive Detection of Prostate and Bladder Cancers

Heena Tyagi [1], Emma Daulton [1], Ayman S. Bannaga [2,3], Ramesh P. Arasaradnam [2,3,4,5] and James A. Covington [1,*]

1 School of Engineering, University of Warwick, Coventry CV4 7AL, UK; Heena.Tyagi@warwick.ac.uk (H.T.); e.daulton@warwick.ac.uk (E.D.)
2 Department of Gastroenterology, University Hospital Coventry & Warwickshire, Coventry CV2 2DX, UK; ayman.bannaga@warwick.ac.uk (A.S.B.); r.arasaradnam@warwick.ac.uk (R.P.A.)
3 Warwick Medical School, University of Warwick, Coventry CV4 7HL, UK
4 School of Health Sciences, Coventry University, Coventry CV1 5FB, UK
5 School of Biological Sciences, University of Leicester, Leicester LE1 7RH, UK
* Correspondence: j.a.covington@warwick.ac.uk

Abstract: Bladder cancer (BCa) and prostate cancer (PCa) are some of the most common cancers in the world. In both BCa and PCa, the diagnosis is often confirmed with an invasive technique that carries a risk to the patient. Consequently, a non-invasive diagnostic approach would be medically desirable and beneficial to the patient. The use of volatile organic compounds (VOCs) for disease diagnosis, including cancer, is a promising research area that could support the diagnosis process. In this study, we investigated the urinary VOC profiles in BCa, PCa patients and non-cancerous controls by using gas chromatography-ion mobility spectrometry (GC-IMS) and gas chromatography time-of-flight mass spectrometry (GC-TOF-MS) to analyse patient samples. GC-IMS separated BCa from PCa (area under the curve: AUC: 0.97 (0.93–1.00)), BCa vs. non-cancerous (AUC: 0.95 (0.90–0.99)) and PCa vs. non-cancerous (AUC: 0.89 (0.83–0.94)) whereas GC-TOF-MS differentiated BCa from PCa (AUC: 0.84 (0.73–0.93)), BCa vs. non-cancerous (AUC: 0.81 (0.70–0.90)) and PCa vs. non-cancerous (AUC: 0.94 (0.90–0.97)). According to our study, a total of 34 biomarkers were found using GC-TOF-MS data, of which 13 VOCs were associated with BCa, seven were associated with PCa, and 14 VOCs were found in the comparison of BCa and PCa.

Keywords: bladder cancer; prostate cancer; urinary biomarkers; urinary VOCs; machine olfaction; GC-IMS, GC-TOF-MS

Citation: Tyagi, H.; Daulton, E.; Bannaga, A.S.; Arasaradnam, R.P.; Covington, J.A. Urinary Volatiles and Chemical Characterisation for the Non-Invasive Detection of Prostate and Bladder Cancers. *Biosensors* **2021**, *11*, 437. https://doi.org/10.3390/bios11110437

Received: 30 September 2021
Accepted: 29 October 2021
Published: 3 November 2021

1. Introduction

Early detection and diagnosis of cancer remains a key goal to improve the prognosis and life expectancy of patients [1–4]. Globally, cancer results in some of the highest mortality rates for any disease. In 2020 alone there were more than 19 million new cancer diagnoses and almost 10 million deaths [5]. The UK is a major contributor to this, with some of the highest cancer rates in the world. It is amongst the top 10% of countries, with the highest number of new cases of cancer [6]. These figures emphasize the importance of using screening methods to improve disease diagnosis and to reduce cancer morbidity [7].

Bladder cancer (BCa) is the ninth most common cancer worldwide and is also one of the most difficult cancers to diagnose and clinically manage [8,9]. Cystoscopy followed by transurethral resection of a bladder tumour (TURBT) with biopsy and histological assessments are considered to be the 'Gold Standard' for the diagnosis of BCa [10]. However, cystoscopy is invasive in nature, and can cause pain, urinary infections, and blood loss in some cases [11,12]. To aid in the diagnosis of BCa, a range of urine tests have been developed including the bladder tumour antigen (BTA) test, nuclear matrix protein 22 (NMP22), urinary bladder cancer antigen (UBC), and fibrin degradation products (FDP). Unfortunately, none of these tests have demonstrated sufficient specificity or sensitivity as a screening test [13].

Prostate cancer (PCa) occurs in men and is the sixth most common cancer worldwide [14–16]. For prostate cancer, PSA (prostate-specific antigen) is a commonly used blood test. However, it lacks sensitivity and specificity. PSA can be used for monitoring PC progression in both symptomatic and asymptomatic patients [17]. The downsides of PSA, as a diagnostic test for PCa patients, are mainly related to the high false-positive rate. PSA can be raised in urinary and prostate infections or other conditions such as benign prostatic hyperplasia (BHP) [18]. Therefore, a raised PSA level can lead to unnecessary biopsies, which may end up causing fever, pain, bleeding, and infection to the patient [19–21]. Recent European Association of Urology guidelines advise undertaking a multiparametric magnetic resonance scan on all patients prior to confirmatory biopsies; however, this is not always accessible, especially in low-resource settings [22].

One area receiving significant interest is in the use of volatile organic compounds (VOCs) to diagnose and monitor cancer. VOCs are chemical compounds that are either produced in vitro or are introduced externally and can indicate the presence, or absence, of disease in the body. The concept first emerged after reports indicated that dogs could recognise cancer by sniffing biological samples [23]. Since this discovery, researchers have reported that VOCs could be used to detect a broad range of cancers including lung, colorectal and pancreatic cancer [24–27].

Urine is a common biological source of VOCs, as the components present are either the intermediate products or end products of metabolic activities occurring inside the human body [28]. A study published in 2016 provided significant evidence for the use of urinary VOCs for distinguishing BCa, from a total of 72 urine samples the results showed an accuracy of 89%, 90% sensitivity, and 88% specificity using PLS-DA (partial least squares discriminant analysis) on GC-MS (gas chromatography-mass spectrometry) data [29].

Urine cytology is a non-invasive test which uses urine as biological modality for the presence of cancer. Several studies showed that though it exhibits high specificity, the sensitivity and specificity highly depend on collection method and cancer grade [30–32].

The gold standard for the analysis of VOCs remains GC-MS, but it is expensive, requiring specialised equipment and trained staff, making it difficult to implement in a point of care scenario. A variant of this is GC-TOF-MS (gas chromatography-time of flight-mass spectrometry), which is a similar technique used for multidimensional analysis of complex samples with the potential to identify an even greater number of VOCs [33,34]. However, more recently a range of other techniques have been reported that have the potential to be used at the point of care. GC-IMS (gas chromatography-ion mobility spectrometry) is one such technique, it provides high sensitivity and selectivity, and the GC-IMS can be created in a portable form factor and can use nitrogen or air as the carrier gas. However, it is less able to identify specific biomarkers and it is unable to identify chemicals with a low proton affinity. Our group has reported the use of this method with a range of different diseases [35,36]. Thus, the combination of GC-TOF-MS, which can provide a means of identifying specific biomarkers, with GC-IMS, a lower cost platform, using air as the carrier and thus facilitating ward use, is advantageous.

The study aimed to identify and test the potential of urinary biomarkers to distinguish between two different cancers and healthy controls using both GC-TOF-MS and GC-IMS. We believe this is the first time that GC-IMS has been used with these cancers in combination with GC-TOF-MS.

2. Materials and Methods

2.1. Urine Samples

A total of 106 patients were recruited after providing informed consent at University Hospital Coventry and Warwickshire NHS Trust, UK. Patients were recruited prior to anti-cancer treatment. This study was approved by Coventry and Warwickshire and North-East Yorkshire NHS Ethics Committees (Ref 18717 and Ref 260179). Urine samples were collected in standard universal sterile specimen containers and frozen within 2 h at −80 °C for subsequent batch analysis and according to standard operating procedures, compliant with tissue bank requirements under Human Tissue Act 2004. No chemicals were added to the urine before freezing, as we have previously shown that urine samples remain stable for extended periods of time at this temperature [37]. Prior to analysis the samples were transferred to the University of Warwick and briefly stored at −20 °C. The samples were defrosted in a laboratory fridge at 4 °C and aliquoted into 20 mL glass sample vials with a crimp cap. We used 5 mL of each urine sample for the analysis with GC-IMS and GC-TOF-MS. Of the 106 urine samples collected, 15 patients had confirmed BCa, 55 were confirmed PCa, and there were 36 non-cancerous controls. The mean age of the BCa patients was 70 years and the mean age of the PCa patients was 72 years. The demographic data of the subjects are illustrated in Table 1.

Table 1. Demographic data for subject groups.

Group	Bladder Cancer	Prostate Cancer		Non-Cancerous
Number of samples	15	55		36
Mean Age (years)	70.0	71.9		62.5
Sex: Male/Female	12:3	All Male		24:12
Mean BMI (Kg/m^{2})	24.4	27.5		30.9
Current Smoker n (% of patients)	1 (6.7%)	6 (10.9%)		3 (8.3%)
Mean PSA level (ng/mL)	-	20.6 (3.6–153.90)		-
Gleason score	-	Case 01	4 + 5 = 9	
		Case 02	3 + 4 = 7	
		Case 03	3 + 3 = 6	
		Case 04	4 + 5 = 9	
		Case 05	4 + 5 = 9	
		Case 06	3 + 4 = 7	
		Case 07	3 + 4 = 7	
		Case 08	3 + 5 = 8	
		Case 09	5 + 4 = 9	
		Case 10	3 + 4 = 7	
		Case 11	3 + 3 = 6	
		Case 12	3 + 4 = 7	
		Case 13	3 + 3 = 6	
		Case 14	4 + 5 = 9	
		Case 15	3 + 4 = 7	
		Case 16	3 + 4 = 7	
		Case 17	3 + 4 = 7	
		Case 18	3 + 4 = 7	
		Case 19	3 + 4 = 7	
		Case 20	3 + 3 = 6	
		Case 21	4 + 5 = 9	
		Case 22	3 + 3 = 6	
		Case 23	4 + 3 = 7	
		Case 24	3 + 4 = 7	
		Case 25	4 + 4 = 8	
		Case 26	3 + 3 = 6	
		Case 27	4 + 5 = 9	
		Case 28	4 + 4 = 8	
		Case 29	3 + 3 = 6	
		Case 30	3 + 3 = 6	
		Case 31	4 + 4 = 8	
		Case 32	3 + 4 = 7	
		Case 33	4 + 5 = 9	
		Case 34	3 + 4 = 7	
		Case 35	3 + 4 = 7	
		Case 36	3 + 4 = 7	

Table 1. *Cont.*

Group	Bladder Cancer	Prostate Cancer	Non-Cancerous
		Case 37 3 + 4 = 7 Case 38 3 + 5 = 8 Case 39 4 + 5 = 9 Case 40 3 + 4 = 7 Case 41 3 + 4 = 7 Case 42 3 + 4 = 7 Case 43 3 + 5 = 8 Case 44 3 + 4 = 7 Case 45 5 + 5 = 10 Case 46 4 + 5 = 9 Case 47 4 + 4 = 8 Case 48 3 + 4 = 7 Case 49 4 + 3 = 7 Case 50 3 + 3 = 6 Case 51 4 + 5 = 9 Case 52 4 + 4 + 8 Case 53 3 + 3 = 6 Case 54 3 + 4 = 7 Case 55 3 + 3 = 6	-
WHO 1973 Grade	Case 01 G2 Case 02 G3 Case 03 G3 Case 04 G1 Case 05 G2 Case 06 G3 Case 07 G1 Case 08 G3 Case 09 G3 Case 10 G1 Case 11 G3 Case 12 G1 Case 13 G1 Case 14 G2 Case 15 G1	-	-

Prostate cancer Gleason grading:
Score ≤ 6, pattern ≤ 3 + 3. This refers to Grade 1. Tumour cells look like normal prostate cells with only individual discrete well-formed glands.
Score 7, pattern 3 + 4. This refers to Grade 2. Tumour with well-form glands and lesser component of poorly differentiated glands.
Score 7, pattern 4 + 3. This refers to Grade 3. Tumour has predominantly poorly formed/fused/cribriform glands with lesser component of well-formed glands.
Score 8, pattern 4 + 4, 3 + 5 and 5 + 3. This refers to Grade 4. Tumour with only poorly formed/fused/cribriform glands.
Score 9 or 10, pattern 4 + 5, 5 + 4 and 5 + 5. This refers to Grade 5. Tumour lacking gland formation (or with necrosis) with or without poorly formed/fused/cribriform glands [38].
G1 low grade differentiation, G2 moderate grade differentiation and G3 is high grade differentiation [39].

2.2. *Analytical Devices*

2.2.1. G.A.S. FlavourSpec Gas Chromatography-Ion Mobility Spectrometry (GC-IMS)

The G.A.S FlavourSpec (Germany) uses a GC-IMS measurement technique to analyse VOCs. GC-IMS is a method used in various applications, such as detection of explosives and chemicals [40–42], air quality [43], health and disease detection [44–46] and food [47–49]. The method is formed of two stages. The first stage is a GC component that pre-separates chemicals based on their interaction with a retentive coating on the inside of a GC column. Thus, chemicals elude from the GC at different times [50]. These chemicals are further analysed using a drift-tube IMS method. Here, the chemicals are ionised (using a tritium source in our case) and pass along a drift-tube, propelled by a high electric field. Against the flow of ions, a buffer gas (using nitrogen in this case) is passed. The buffer gas and the ions collide resulting in a loss of momentum of the ions. Thus, the transit time along the tube is a function of the interaction of the ion with the electric field and the number of collisions with the buffer gas. This provides two-dimensional separation of the chemical components [48,51].

For analysis, glass vials containing samples were transferred to an autosampler fitted to the GC-IMS. The sample tray was chilled to 4 °C to reduce sample degradation during sample analysis. Each sample was heated to 40 °C and agitated for 10 min before sampling. The autosampler then took 0.5 mL of sample headspace and directly injected it into the GC-IMS. Urinary headspace was defined as the volume of gas above the urine sample inside the vial, which was in chemical equilibrium with liquid phase urine. The machine settings for analysis were as follows: E1: 150 mL/min (for the drift tube IMS), E2: 20 mL/min (for the GC column), and the pump was set to 25%. The total run time per sample was 10 min. The temperatures were set to T1 (IMS): 45 °C, T2 (column): 80 °C, and T3 (injector): 70 °C.

2.2.2. Markes Gas Chromatography Time-of-Flight Mass Spectrometry (GC-TOF-MS)

GC-TOF-MS operates by analysing the time of flight of ions and analyse them according to their mass-to-charge ratio. The GC-TOF-MS system used was a combination of a TRACE 1300 GC (Thermo Fisher Scientific, Loughborough, UK) and a BenchTOF-HD TOF-MS (Markes Intl., Llantrisant, UK). This system also included a high-throughput autosampler and a thermal desorption unit, ULTRA-xr and UNITY-xr, respectively (both from Markes Intl.). The GC separated the chemicals in the same way as explained previously. The separated chemicals were detected by TOF MS once they entered the TOF 'flight box'. TOF-MS separates fragment ions instead of molecular ions as in an IMS. The ions are detected depending upon the mass-to-charge ratio of the ions after passing through the drift tube [52,53].

For analysis, a thermal desorption (TD) sorbent tube (C2-AXXX-5149, Markes Intl., Llantrisant, UK) was inserted through the septum and into the headspace above the sample and then heated at 40 °C for 20 min. A pump was then attached to the TD tube, and whilst still being heated to 40 °C, the headspace VOCs were then pulled onto the tubes at 20 mL/minute for a further 20 min. The sorbent tubes were then placed in an autosampler for analysis. The analysis started with ULTRA-xr with a stand-by split set to 150 °C. The GC run time for samples was 25 min with a programmed temperature ramp from 40 °C to 280 °C at 20 °C/min. Each sample was pre-purged for 1 min and then desorbed at 250 °C for 10 min, with the trap purge time set to 1 min. These traps were then cooled at −30 °C and the trap was purged for 3 min at a temperature of 300 °C. The temperature for both transfer line and ion source were heated to 250 °C. The chemicals from GC-TOF-MS analysis were identified using the national institute of standards and technology (NIST) list (2011).

2.3. *Statistical Methods*

For GC-IMS data analysis, the data were extracted using the G.A.S VOCal (v0.1.3, G.A.S., Dortmund, Germany) software. This was followed by pre-processing steps to reduce the data's dimensionality. Among all the data points, the central section contained all the computationally significant chemical information and thus all the other data were removed through a cropping process. This was followed by applying a small threshold to remove the background information, which was a value just above the background noise level. The same data cropping and threshold values were used on all the data, and it was undertaken using an automated program. The data were then analysed using a 10-fold cross-validation, undertaken using a bespoke R program (version 3.6.2). Within each fold training set, feature selection was undertaken using a Wilcoxon rank-sum test between the different cancer groups and non-cancerous group. That resulted in the identification of the 20 most discriminatory features between the two groups and the features trained by three models, XGBoost, logistic regression, and random forest. The model was then applied to the test set to create class probabilities. Once all the samples had been within a test set, statistical results were generated from the probabilities, including a receiver operator characteristic (ROC) curve, area under the curve (AUC), sensitivity, specificity, positive predictive value (PPV), and negative predictive value (NPV).

An analogous approach was used for GC-TOF-MS data analysis. For GC-TOF-MS, the chemicals and the abundance of the chemicals were identified. Using the TOF-DS software, a background correction was applied, and the chromatogram was integrated, and the peaks were identified using the NIST list which was exported. The data obtained from GC-TOF-MS were converted into text files of chemical lists and abundances. The data were then processed using an 'R' program that was like that used for GC-IMS, where chemical components of discriminative power were identified. Figure 1 provides a flow diagram of the data analysis steps.

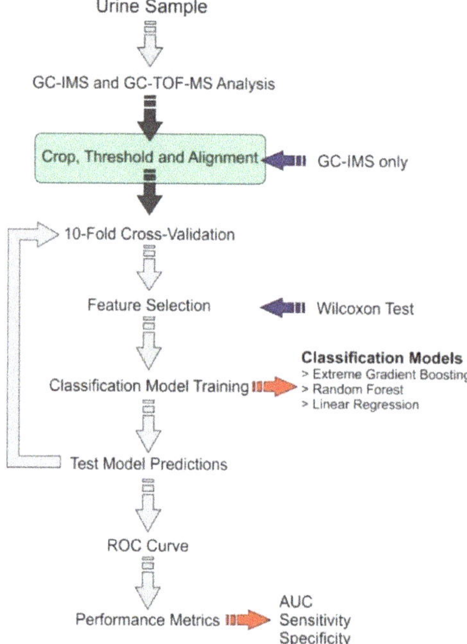

Figure 1. Data analysis pipeline.

3. Results

Figure 2 shows a typical output from the GC-IMS method from a urine sample in which the x-axis represents the drift time of the IMS and the y-axis represents the retention time of the GC. In the figure, the 'dots' are the chemicals detected by the IMS and the intensity of the peak represents the number of ions. Those 'dots' in red are the most intense. The red line in the figure is the default output of the instrument where no chemicals are present. G.A.S VOCal (v0.1.3, G.A.S., Dortmund, Germany) was used to view the GC-IMS data.

Figure 3 provides an example output from the GC-TOF-MS method. Here, the x-axis refers to the retention time, and the y-axis, the total ion count.

The results of the statistical analysis of the GC-IMS gathered results between different cancer groups and the non-cancerous group are given in Table 2. The results demonstrate high sensitivity and specificity, indicating that there are significant differences between the VOC profiles of the different groups. Importantly, good separation between the two different cancers, BCa vs. PCa, was also achieved. The false negative rate calculated for the GC-IMS analytical method in the study was 0.40 for BCa versus the PCa comparison,

0.13 for BCa versus the non-cancerous group and 0.24 for PCa versus the non-cancerous group, whereas the false positive rate was 0.02 for BCa versus the PCa group, 0.08 for BCa versus the non-cancerous group and 0.12 for PCa versus the non-cancerous group.

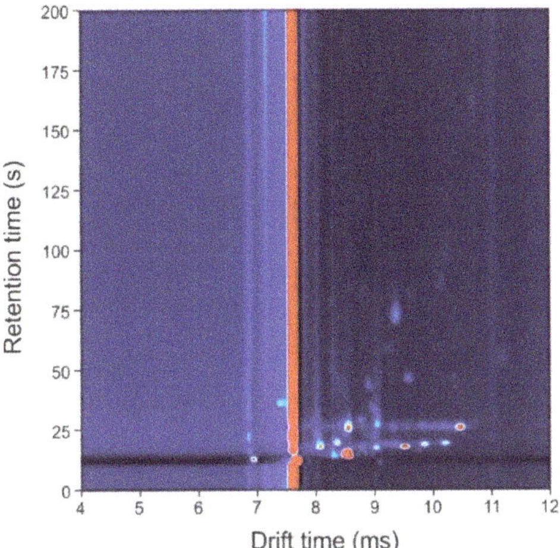

Figure 2. Typical output plot from the gas chromatography-ion mobility spectrometry (GC-IMS) instrument.

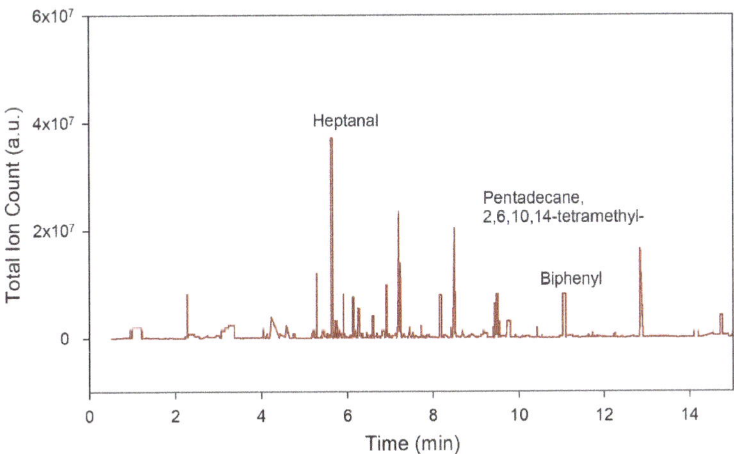

Figure 3. Figure illustrates a typical output plot of gas chromatography time-of-flight mass spectrometry (GC-TOF-MS). The x-axis in the plot represents the retention time and y-axis lists the chemical according to their abundance in the sample.

Table 2. GC-IMS diagnostic group results.

Comparisons	Classifiers	AUC	Sensitivity	Specificity	PPV	NPV
BCa vs. PCa	Logistic Regression with Elastic Net Regularization	0.97 (0.93–1.00)	0.60 (0.38–0.80)	0.98 (0.95–1.00)	0.90	0.90
BCa vs. non-Cancerous	Logistic Regression with Elastic Net Regularization	0.95 (0.90–0.99)	0.87 (0.70–1.00)	0.92 (0.84–0.98)	0.81	0.95
PCa vs. non-Cancerous	Extreme Gradient Boosting	0.89 (0.83–0.94)	0.76 (0.64–0.88)	0.88 (0.80–0.95)	0.81	0.85

The ROC curves obtained from GC-IMS data comparing BCa and the non-cancerous group, BCa and PCa groups, and PCa and non-cancerous groups are shown in Figure 4. The results indicate that among BCa patients and PCa patients, AUC (area under the curve) was 0.97 (0.93–1.00) with sensitivity and specificity of 0.60 (0.38–0.80) and 0.98 (0.95–1.00), respectively. However, the separation between BCa and non-cancerous samples was even higher with a sensitivity of 0.87 (0.70–1.00), specificity of 0.92 (0.84–0.98) and AUC of 0.95 (0.90–0.99). Similarly, for PCa vs non-cancerous samples using GC-IMS, the separation was significant with a sensitivity of 0.76 (0.64–0.88), specificity of 0.88 (0.80–0.95) and AUC of 0.89 (0.83–0.94).

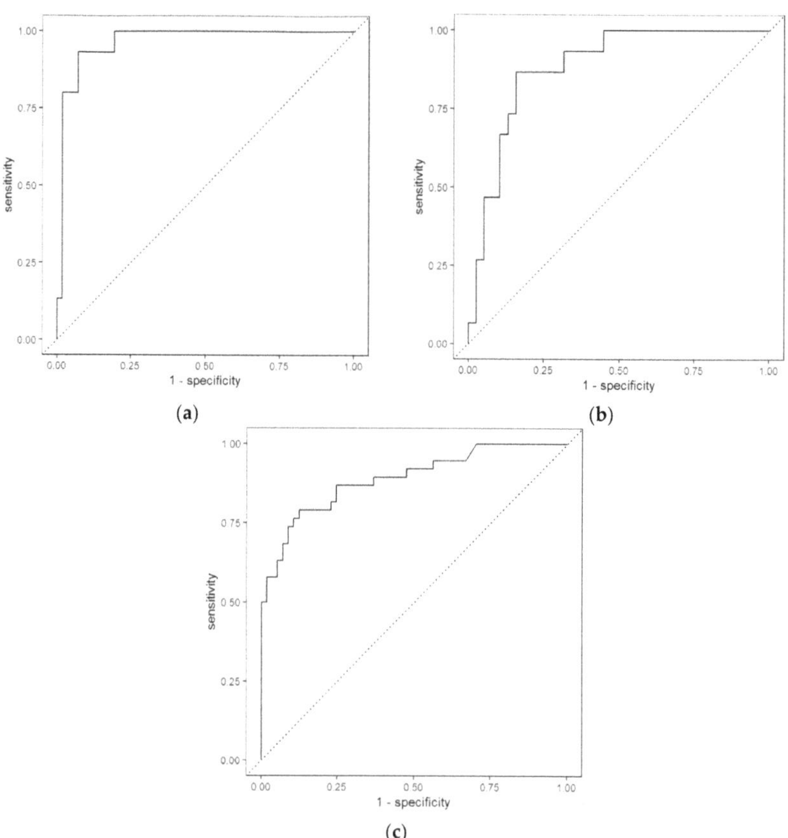

Figure 4. Receiver operator characteristic for (**a**) bladder cancer vs. PCa; (**b**) bladder cancer vs. non-cancerous group; and (**c**) prostate cancer vs. non-cancerous group using GC-IMS.

The results of the statistical analysis between different cancer groups for GC-TOF-IMS are given in Table 3. The results demonstrate high sensitivity and specificity, indicating that there are significant differences between the VOC profiles of different cancer groups, which was also shown in the GC-IMS data. The results showed that the false negative rate for BCa versus PCa comparison was 0.47, for BCa versus the non-cancerous group was 0.73 and for PCa versus the non-cancerous group was 0.22 for the GC-TOF-MS analytical method. The false positive rate for BCa versus PCa comparison was 0.1, for BCa versus the non-cancerous group it was 0.06, and for PCa versus the non-cancerous group it was 0.12.

Table 3. GC-TOF-MS diagnostic group results.

Comparisons	Classifiers	AUC	Sensitivity	Specificity	PPV	NPV
BCa vs. PCa	Logistic Regression with Elastic Net Regularization	0.84 (0.73–0.93)	0.53 (0.33–0.75)	0.90 (0.83–0.96)	0.62	0.87
BCa vs. non-Cancerous	Random Forest	0.81 (0.70–0.90)	0.27 (0.09–0.46)	0.94 (0.88–1.00)	0.33	0.71
PCa vs. Non-Cancerous	Random Forest	0.94 (0.90–0.97)	0.78 (0.66–0.89)	0.88 (0.80–0.95)	0.82	0.85

The ROC curves obtained from GC-TOF-MS data comparing BCa and non-cancerous groups, BCa and PCa groups, and PCa and non-cancerous groups are shown in Figure 5. The results indicate that GC-TOF-MS was able to differentiate BCa and PCa with AUC 0.84 (0.73–0.93), sensitivity and specificity of 0.53 (0.33–0.75) and 0.90 (0.83–0.96). The separation between BCa and non-cancerous samples was very poor with sensitivity only 0.27 (0.9–0.46), specificity 0.94 (0.88–1.00) and AUC 0.82 (0.72–0.90). However, the separation was more significant with sensitivity 0.78 (0.66–0.89), specificity 0.88 (0.80–0.95) and AUC 0.94 (0.90–0.97) for PCa and non-cancerous groups.

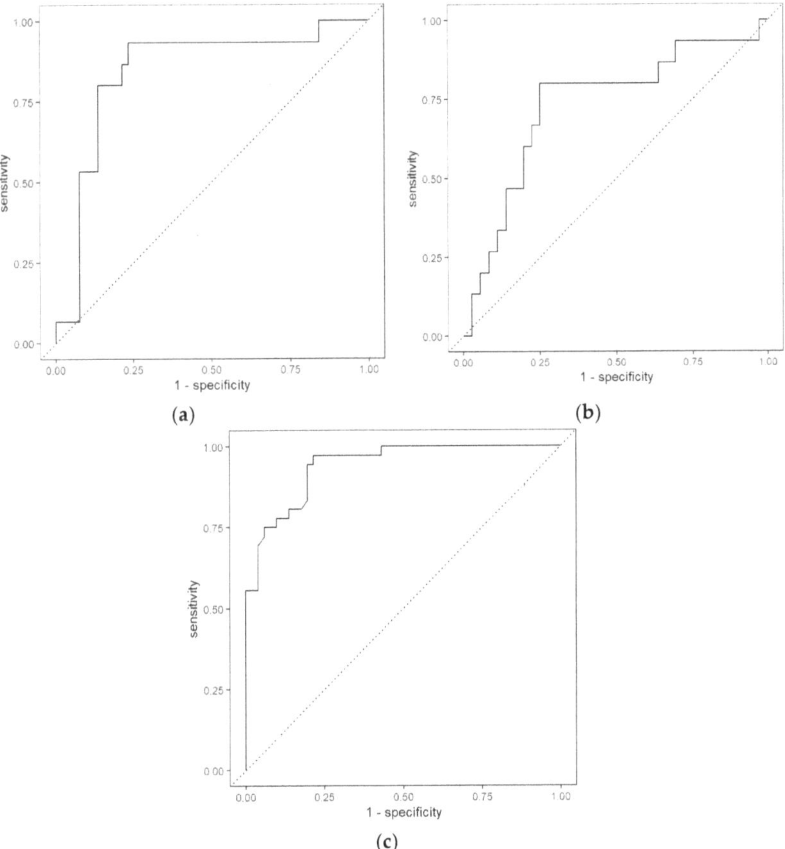

Figure 5. ROC for (**a**) bladder cancer vs. PCa; (**b**) bladder cancer vs. non-cancerous group; and (**c**) prostate cancer vs. non-cancerous group using GC-TOF-MS.

In our results, we analysed different VOCs linked to BCa and PCa for the screening and diagnosis of these cancers. A total of 34 biomarkers were found using TOF-DS software. These VOCs were verified using PubChem, NIST (National Institute of Standards and Technology), and previously published papers. Out of 34, 13 VOCs were found in the comparison of BCa and non-cancerous groups specific to BCa, as shown in Table 4, seven in PCa and non-cancerous groups specific to PCa, as shown in Table 5, and 14 VOCs were found in the comparison of BCa and PCa group, as shown in Table 6, out of which 3 VOCs do not overlap either with BCa or PCa, which may indicate that they are new markers.

Table 4. A list of possible biomarkers from the analysis of urine samples by GC-TOF-MS identified using PubChem, NIST and publications significant to bladder cancer.

	Chemicals	*p*-Values	Molecular Weight (g/mol)
1	Biphenyl	<0.01	154.21
2	Nonanal	<0.01	142.24
3	Tetradecane	<0.01	198.39
4	Pentadecane, 2,6,10,14-tetramethyl-	0.012	268.5
5	2-Pentanone	0.012	86.13
6	Undecane	0.014	156.31
7	4-Heptanone	0.018	114.19
8	Dodecane	0.025	170.33
9	Hexadecane	0.026	226.44
10	Heptanal	0.026	114.19
11	Methyl Isobutyl Ketone	0.045	100.16
12	Naphthalene	0.046	128.169
13	Benzoic acid	0.049	122.12

Table 5. List of possible biomarkers from the analysis of urine samples by GC-TOF-MS identified using PubChem, NIST and publications significant to PCa.

	Chemicals	*p*-Values	Molecular Weight (g/mol)
1	Toluene	<0.01	92.14
2	Phenol	<0.01	325.4
3	Acetic acid	<0.01	60.05
4	1-Hexanol, 2-ethyl-	0.011	130.229
5	Disulfide, dimethyl	0.012	94.2
6	Cyclopentanone, 2-methyl-	0.017	98.14
7	Pyrrole	0.033	67.09

Table 6. List of possible biomarkers from the analysis of urine samples by GC-TOF-MS identified using PubChem, NIST and publications significant to PCa and bladder cancer.

	Chemicals	*p*-Values	Molecular Weight (g/mol)
1	Toluene	<0.01	92.14
2	Methyl Isobutyl Ketone	<0.01	100.16
3	Dodecane	<0.01	170.33
4	Phenol	<0.01	325.4
5	Cyclopentanone, 2-methyl-	<0.01	98.14
6	2-Hexanone	<0.01	100.16
7	Heptanal	<0.01	114.19
8	p-Xylene	<0.01	106.16
9	Nonane, 3-methyl-	<0.01	142.28
10	Tetradecane	<0.01	198.39
11	Nonanal	<0.01	142.24
12	Biphenyl	0.019	154.21
13	Acetic acid	0.025	60.05
14	2-Pentanone	0.032	86.13

4. Discussion

In our study, we found that both GC-IMS and GC-TOF-MS were able to separate different cancer groups from each other as well as non-cancerous group. The separation between BCa from non-cancerous group was highest using GC-IMS with 0.95 AUC (0.87 sensitivity and 0.92 specificity). A similar study conducted by Weber et al. [54] suggested overall accuracy of 70% (70% sensitivity and 70% specificity) using urinary headspace for the analysis of BCa using gas sensors. Another study conducted by Khalid et al. [55] showed very high statistical results using an in-house GC-sensor device. They used two models for analysis suggesting 100% sensitivity and 94.6% specificity using a linear discriminant analysis (LDA) model and 95.8% sensitivity and 94.6% specificity using PLS-DA.

The separation between PCa and the non-cancerous group was highest using GC-TOF-MS method with 0.94 AUC (0.78 sensitivity and 0.88 specificity) whereas the study conducted by Gao et al. [56] for the analysis of urinary VOCs for prostate cancer calculated 0.92 AUC (0.96 sensitivity and 0.80 specificity). Another study conducted by Lima et al. [57] used PLS-DA to discriminate PCa from non-cancerous group with an AUC of 0.83 (84% sensitivity and 80% specificity) using urine headspace.

In this study, we developed urinary VOC profiles linked with BCa and PCa. Table 4 consists of the chemicals that have been identified in our study and have been cross verified using PubChem, NIST and previously published research, which may have relevance to BCa diagnosis.

Out of 13 VOCs found to be noteworthy to BCa, biphenyl, heptanal, and 2,6,10, 14-tetramethyl-pentadecane were the three distinct biomarkers found in our study that did not overlap with other studies. Biphenyl has been identified as the most significant biomarker in our study. Biphenyl has been linked to various diseases, including carcinoma. It has been proven that biphenyl is a promoter of BCa in rats [58]. Biphenyl has been found to be metabolized in the liver [59]. Heptanal is reported to present in the blood of lung cancer patients [60]. According to the HMBD (Human Metabolome Database), the biological activity of heptanal inside humans can cause digestive disorder including associated with the bladder [61]. 2,6,10,14-tetramethyl-pentadecane is reported as carcinogens but is mentioned far less in the literature [62]. Nonanal, tetradecane, dodecane, hexadecane, naphthalene, and methyl isobutyl ketone were suggested by Rodrigues et al. [63] in their study using GC-MS on BCa cell lines whereas 2-pentanone and 4-heptanone overlap with the findings of Cauchi et al. [29]. Benzoic acid was another chemical found in our study that overlapped in both Rodrigues et al. [63] and Cauchi et al. [29].

From the analysis of PCa urine samples, a total of seven distinct VOCs were identified and are summarised in Table 5. In our study, we found toluene as the most significant chemical for PCa. Toluene has been published previously as a significant biomarker for PCa [64]. In addition, it has been reported that toluene has been found to be associated with testicular diseases [65,66]. Pyrrole has been reported by Smith et al. in their study with 24 controls and 13 patients with PCa. They tested the urine samples to assess VOC profiles and found pyrrole to be one of the significant markers for PCa [67]. 2-Ethyl-1-hexanol, phenol and dimethyl disulphide [68], acetic acid [69], and 2-methyl cyclopentanone [70] were also found in our study, which overlaps with previous studies.

Table 6 represents all the chemicals found in the analysis of urine samples for prostate versus BCa. Most of the chemicals present in this list are like those found in Tables 4 and 5. 2-Hexanone, p-xylene, and 3-methyl nonane are the only significant chemicals out of 14 in this list that are important for separating BCa and PCa. 2-Hexanone and p-xylene have previously been reported as significant markers for the PCa [68,70]. There is no significant evidence for both 2-hexanone and p-xylene as a potential biomarker for BCa. However, 3-methyl-Nonane has not yet been reported as a biomarker for either bladder or PCa, although they have been reported as a biomarker for lung cancer in different studies [71,72]. This may signify the importance of 3-methyl-nonane as a potentially significant marker. The results reported in this paper support the findings of other groups for the validation of these chemicals as potential biomarkers in both PCa and BCa. It has been noted that the chemicals found in all the cancer groups were different and there was almost no overlapping of the VOC fingerprints for BCa and PCa. This adds further support to the unique VOC fingerprint in cancers of different cell origins [73].

The use of urinary VOC analysis is an attractive option due to the non-invasive nature. It also has the potential to be used in early cancer diagnosis with further validation studies. This approach may also prove to be efficient, whilst lowering the cost per patient, and increasing patient compliance due to its non-invasive nature. The results of using GC-IMS as an analysis tool are significant as the method is much simpler than using a high-end analytical method, such as GC-MS, and without the need for a laboratory environment. We believe that using VOCs to analyse human waste will be an important diagnostic tool for the future. Cancer may well be one area of focus and may be used as part of the UK 2-week wait screening program to help reduce the number of unneeded procedures. The key is to run more larger studies targeting these cancers and to have tools that are CE marked (or equivalent) for cancer diagnosis. We plan to use urine VOCs in association with other tests in future which help to improve the performance and achieve a more in depth understanding of VOCs and their metabolic pathways.

Our results were limited by not accounting for the contributory factors that can also lead to abnormal metabolism with subsequent excretion of differing concentrations of these chemicals in the urine. These factors include stress, alcohol, smoking, certain food products, medicines, and different environmental factors. Several studies have reported the effect of smoking on VOCs [74,75]. Study conducted by A. McWilliams et al. showed that active smoking had an impact on urinary VOC profiles associated with current smokers and ex-smokers [76]. We aim to consider these further in the next study. We also did not undertake full chemical identification with calibration standards. However, many of the chemicals we found correlate with other studies and, therefore, there is evidence that these are correct.

5. Conclusions

In this paper, GC-IMS and GC-TOF-MS methods were used to identify VOC fingerprints using urine headspace and establish an interdependence between BCa, PCa and non-cancerous samples. It was found that both GC-IMS and GC-TOF-MS have the potential to differentiate between different cancer groups with respective AUC for different diagnostic groups: For GC-IMS, BCa and PCa (0.97 (0.93–1)), BCa and non-cancerous (0.95 (0.90–0.99)), PCa and non-cancerous (0.89 (0.83–0.94)) and for GC-TOF-MS, BCa and PCa (0.84 (0.73–0.93)), BCa and non-cancerous (0.81(0.70–0.90)), PCa and non-cancerous (0.94 (0.90–0.97)). A total of 35 VOCs were found to be relevant for identifying these cancer groups, with several VOCs distinct to each cancer. VOCs from this study were supported by findings from previous studies. This signifies that VOCs for both bladder and prostate cancer have different profiles, which may be helpful in future to distinguish them. In the future, the VOC profiles obtained from these analytical devices can be used as a reference for developing low-cost devices. It is plausible that VOC profiles can be used as an adjunct to diagnosis enabling selection of only high-risk groups to undergo cystoscopy examinations which will be widely beneficial considering limited capacity and cost.

Author Contributions: Conceptualization, H.T., E.D. and J.A.C.; methodology, H.T. and E.D.; formal analysis, H.T.; investigation, H.T.; resources, J.A.C. and R.P.A.; data curation, H.T. and A.S.B.; writing—original draft preparation, H.T.; writing—review and editing, H.T., E.D., A.S.B. and J.A.C. and R.P.A.; visualization, H.T. and J.A.C.; supervision, J.A.C.; project administration, J.A.C. and R.P.A. All authors have read and agreed to the published version of the manuscript.

Funding: This research received no external funding.

Institutional Review Board Statement: This study was approved by Coventry and Warwickshire and North-East Yorkshire NHS Ethics Committees (Ref 18717 and Ref 260179).

Informed Consent Statement: Informed consent was obtained from all subjects involved in the study. Written informed consent has been obtained from the patient(s) to publish this paper.

Data Availability Statement: The data presented in this study are available on request from the corresponding author. The data are not publicly available due to the ethical approval.

Conflicts of Interest: The authors declare no conflict of interest.

References

1. Ott, J.J.; Ullrich, A.; Miller, A.B. The importance of early symptom recognition in the context of early detection and cancer survival. *Eur. J. Cancer* **2009**, *45*, 2743–2748. [CrossRef]
2. de Nooijer, J.; Lechner, L.; de Vries, H. Early detection of cancer: Knowledge and behavior among Dutch adults. *Cancer Detect. Prev.* **2002**, *26*, 362–369. [CrossRef]
3. Bloom, J.R. Early detection of cancer. Psychologic and social dimensions. *Cancer* **1994**, *74*, 1464–1473. [CrossRef]
4. McPhail, S.D.; Johnson, S.T.; Greenberg, D.S.; Peake, M.D.; Rous, B. Stage at diagnosis and early mortality from cancer in England. *Br. J. Cancer* **2015**, *112*, S108–S115. [CrossRef]
5. Sung, H.; Ferlay, J.; Siegel, R.L.; Laversanne, M.; Soerjomataram, I.; Jemal, A.; Bray, F. Global cancer statistics 2020: GLOBOCAN estimates of incidence and mortality worldwide for 36 cancers in 185 countries. *CA A Cancer J. Clin.* **2021**, *71*, 209–249. [CrossRef] [PubMed]
6. Bray, F.; Ferlay, J.; Soerjomataram, I.; Siegel, R.L.; Torre, L.A.; Jemal, A.E. Global cancer statistics 2018: GLOBOCAN estimates of incidence and mortality worldwide for 36 cancers in 185 countries. *CA A Cancer J. Clin.* **2018**, *68*, 394–424. [CrossRef] [PubMed]
7. Richards, M.A. The size of the prize for earlier diagnosis of cancer in England. *Br. J. Cancer* **2009**, *101*, S125–S129. [CrossRef]
8. Saginala, K.; Barsouk, A.; Aluru, J.S.; Rawla, P.; Padala, S.A.; Barsouk, A. Epidemiology of Bladder Cancer. *Med. Sci.* **2020**, *8*, 15. [CrossRef]
9. Richters, A.; Aben, K.K.H.; Kiemeney, L.A.L.M. The global burden of urinary bladder cancer: An update. *World J. Urol.* **2019**, *38*, 1895–1904. [CrossRef]
10. Yafi, F.A.; Brimo, F.; Steinberg, J.; Aprikian, A.G.; Tanguay, S.; Kassouf, W. Prospective analysis of sensitivity and specificity of urinary cytology and other urinary biomarkers for bladder cancer. *Urol. Oncol. Semin. Orig. Investig.* **2015**, *33*, 66.e25–66.e31. [CrossRef]
11. Planz, B.; Jochims, E.; Deix, T.; Caspers, H.P.; Jakse, G.; Boecking, A. The role of urinary cytology for detection of bladder cancer. *Eur. J. Surg. Oncol. (EJSO)* **2005**, *31*, 304–308. [CrossRef]
12. Kumar, N.; Talwar, R.; Nandy, P.R. Efficacy of voided urinary cytology and ultrasonography compared to cystoscopy in the detection of urinary bladder cancer. *Afr. J. Urol.* **2017**, *23*, 192–196. [CrossRef]
13. Budman, L.I.; Kassouf, W.; Steinberg, J.R. Biomarkers for detection and surveillance of bladder cancer. *Can. Urol. Assoc. J.* **2013**, *2*, 212–221. [CrossRef] [PubMed]
14. Culp, M.B.; Soerjomataram, I.; Efstathiou, J.A.; Bray, F.; Jemal, A. Recent Global Patterns in Prostate Cancer Incidence and Mortality Rates. *Eur. Urol.* **2019**, *77*, 38–52. [CrossRef] [PubMed]
15. Bleyer, A.; Spreafico, F.; Barr, R. Prostate cancer in young men: An emerging young adult and older adolescent challenge. *Cancer* **2020**, *126*, 46–57. [CrossRef]
16. Deev, V.; Solovieva, S.; Andreev, E.; Protoshchak, V.; Karpushchenko, E.; Sleptsov, A.; Kartsova, L.; Bessonova, E.; Legin, A.; Kirsanov, D. Prostate cancer screening using chemometric processing of GC–MS profiles obtained in the headspace above urine samples. *J. Chromatogr. B* **2020**, *1155*, 122298. [CrossRef]
17. van den Bergh, R.C.N.; Roemeling, S.; Roobol, M.J.; Roobol, W.; Schröder, F.H.; Bangma, C.H. Prospective Validation of Active Surveillance in Prostate Cancer: The PRIAS Study. *Eur. Urol.* **2007**, *52*, 1560–1563. [CrossRef]
18. Catalona, W.J.; Smith, D.S.; Ratliff, T.L.; Dodds, K.M.; Coplen, D.E.; Yuan, J.J.; Petros, J.A.; Andriole, G.L. Measurement of Prostate-Specific Antigen in Serum as a Screening Test for Prostate Cancer. *N. Engl. J. Med.* **1991**, *324*, 1156–1161. [CrossRef]
19. Schröder, F.H.; Hugosson, J.; Roobol, M.J.; Tammela, T.L.J.; Ciatto, S.; Nelen, V.; Kwiatkowski, M.; Lujan, M.; Lilja, H.; Zappa, M.; et al. Screening and Prostate-Cancer Mortality in a Randomized European Study. *N. Engl. J. Med.* **2009**, *360*, 1320–1328. [CrossRef] [PubMed]
20. Rosario, D.J.; Lane, J.A.; Metcalfe, C.; Donovan, J.L.; Doble, A.; Goodwin, L.; Davis, M.; Catto, J.W.F.; Avery, K.; Neal, D.E.; et al. Short term outcomes of prostate biopsy in men tested for cancer by prostate specific antigen: Prospective evaluation within ProtecT study. *BMJ* **2012**, *344*, d7894. [CrossRef]
21. Wade, J.; Rosario, D.J.; Macefield, R.C.; Avery, K.N.; Salter, C.E.; Goodwin, M.L.; Blazeby, J.M.; Lane, J.A.; Metcalfe, C.; Neal, D.E. Psychological impact of prostate biopsy: Physical symptoms, anxiety, and depression. *J. Clin. Oncol.* **2013**, *31*, 4235–4241. [CrossRef]
22. Mottet, N.; Bellmunt, J.; Bolla, M.; Briers, E.; Cumberbatch, M.G.; De Santis, M.; Fossati, N.; Gross, T.; Henry, A.M.; Joniau, S.; et al. EAU-ESTRO-SIOG Guidelines on Prostate Cancer. Part 1: Screening, Diagnosis, and Local Treatment with Curative Intent. *Eur. Urol.* **2017**, *71*, 618–629. [CrossRef]
23. Williams, H.; Pembroke, A. Sniffer dogs in the melanoma clinic? *Lancet* **1989**, *333*, 734. [CrossRef]
24. Chang, J.-E.; Lee, D.-S.; Ban, S.-W.; Oh, J.; Jung, M.Y.; Kim, S.-H.; Park, S.; Persaud, K.; Jheon, S. Analysis of volatile organic compounds in exhaled breath for lung cancer diagnosis using a sensor system. *Sensors Actuators B Chem.* **2018**, *255*, 800–807. [CrossRef]

25. Bond, A.; Greenwood, R.; Lewis, S.; Corfe, B.; Sarkar, S.; O'Toole, P.; Rooney, P.; Burkitt, M.; Hold, G.; Probert, C. Volatile organic compounds emitted from faeces as a biomarker for colorectal cancer. *Aliment. Pharmacol. Ther.* **2019**, *49*, 1005–1012. [CrossRef]
26. Arasaradnam, R.P.; Wicaksono, A.; O'Brien, H.; Kocher, H.M.; Covington, J.A.; Crnogorac-Jurcevic, T. Noninvasive diagnosis of pancreatic cancer through detection of volatile organic compounds in urine. *Gastroenterology* **2018**, *154*, 485–487. [CrossRef] [PubMed]
27. Staff, P.O. Correction: Detection of Colorectal Cancer (CRC) by Urinary Volatile Organic Compound Analysis. *PLoS ONE* **2015**, *10*, e0118975.
28. Becker, R. Non-invasive cancer detection using volatile biomarkers: Is urine superior to breath? *Med. Hypotheses* **2020**, *143*, 110060. [CrossRef]
29. Cauchi, M.; Weber, C.M.; Bolt, B.J.; Spratt, P.B.; Bessant, C.; Turner, D.C.; Willis, C.M.; Britton, L.E.; Turner, C.; Morgan, G. Evaluation of gas chromatography mass spectrometry and pattern recognition for the identification of bladder cancer from urine headspace. *Anal. Methods* **2016**, *8*, 4037–4046. [CrossRef]
30. Renshaw, A.A.; Nappi, D.; Weinberg, D.S. Cytology of Grade 1 Papillary Transitional Cell Carcinoma. *Acta Cytol.* **1996**, *40*, 676–682. [CrossRef] [PubMed]
31. Raab, S.S.; Grzybicki, D.M.; Vrbin, C.M.; Geisinger, K.R. Urine cytology discrepancies: Frequency, causes, and outcomes. *Am. J. Clin. Pathol.* **2007**, *127*, 946–953. [CrossRef]
32. Muus Ubago, J.; Mehta, V.; Wojcik, E.M.; Barkan, G.A. Evaluation of atypical urine cytology progression to malignancy. *Cancer Cytopathol.* **2013**, *121*, 387–391. [CrossRef]
33. Chang, K.L.; Ho, P.C. Gas Chromatography Time-Of-Flight Mass Spectrometry (GC-TOF-MS)-Based Metabolomics for Comparison of Caffeinated and Decaffeinated Coffee and Its Implications for Alzheimer's Disease. *PLoS ONE* **2014**, *9*, e104621. [CrossRef]
34. Rudnicka, J.; Kowalkowski, T.; Ligor, T.; Buszewski, B. Determination of volatile organic compounds as biomarkers of lung cancer by SPME–GC–TOF/MS and chemometrics. *J. Chromatogr. B* **2011**, *879*, 3360–3366. [CrossRef]
35. Tiele, A.; Wicaksono, A.; Daulton, E.; Ifeachor, E.; Eyre, V.; Clarke, S.; Timings, L.; Pearson, S.; Covington, J.A.; Li, X. Breath-based non-invasive diagnosis of Alzheimer's disease: A pilot study. *J. Breath Res.* **2020**, *14*, 26003. [CrossRef]
36. Bosch, S.; Bot, R.; Wicaksono, A.; Savelkoul, E.; van der Hulst, R.; Kuijvenhoven, J.; Stokkers, P.; Daulton, E.; Covington, J.A.; de Meij, T.G.J.; et al. Early detection and follow-up of colorectal neoplasia based on faecal volatile organic compounds. *Colorectal Dis.* **2020**, *22*, 1119–1129. [CrossRef]
37. Esfahani, S.; Sagar, N.M.; Kyrou, I.; Mozdiak, E.; O'Connell, N.; Nwokolo, C.; Bardhan, K.D.; Arasaradnam, R.P.; Covington, J.A. Variation in Gas and Volatile Compound Emissions from Human Urine as It Ages, Measured by an Electronic Nose. *Biosensors* **2016**, *6*, 4. [CrossRef] [PubMed]
38. Epstein, J.I.; Egevad, L.; Amin, M.B.; Delahunt, B.; Srigley, J.R.; Humphrey, P.A. The 2014 International Society of Urological Pathology (ISUP) consensus conference on Gleason grading of prostatic carcinoma. *Am. J. Surg. Pathol.* **2016**, *40*, 244–252. [CrossRef] [PubMed]
39. Mostofi, F.K. Histological typing of urinary bladder tumors. *Int. Histol. Classif. Tumors* **1973**, *17*. Available online: https://apps.who.int/iris/handle/10665/41533 (accessed on 16 October 2021).
40. Haley, L.V.; Romeskie, J.M. Development of an explosives detection system using fast GC-IMS technology. In Proceedings of the IEEE 32nd Annual 1998 International Carnahan Conference on Security Technology (Cat. No.98CH36209), Alexandria, VA, USA, 12–14 October 1998; pp. 59–64. [CrossRef]
41. Cook, G.W.; LaPuma, P.T.; Hook, G.L.; Eckenrode, B.A. Using Gas Chromatography with Ion Mobility Spectrometry to Resolve Explosive Compounds in the Presence of Interferents*. *J. Forensic Sci.* **2010**, *55*, 1582–1591. [CrossRef]
42. Kwan, C.; Snyder, A.P.; Erickson, R.P.; Smith, P.A.; Maswadeh, W.M.; Ayhan, B.; Jensen, J.L.; Jensen, J.O.; Tripathi, A. Chemical Agent Detection Using GC-IMS: A Comparative Study. *IEEE Sensors J.* **2010**, *10*, 451–460. [CrossRef]
43. Moura, P.C.; Vassilenko, V.; Fernandes, J.M.; Santos, P.H. Indoor and Outdoor Air Profiling with GC-IMS. In *Doctoral Conference on Computing, Electrical and Industrial Systems*; Springer: Berlin/Heidelberg, Germany, 2020.
44. Allers, M.; Langejuergen, J.; Gaida, A.; Holz, O.; Schuchardt, S.; Hohlfeld, J.M.; Zimmermann, S. Measurement of exhaled volatile organic compounds from patients with chronic obstructive pulmonary disease (COPD) using closed gas loop GC-IMS and GC-APCI-MS. *J. Breath Res.* **2016**, *10*, 26004. [CrossRef]
45. Tiele, A.; Wicaksono, A.; Kansara, J.; Arasaradnam, R.P.; Covington, J.A. Breath Analysis Using eNose and Ion Mobility Technology to Diagnose Inflammatory Bowel Disease—A Pilot Study. *Biosensors* **2019**, *9*, 55. [CrossRef] [PubMed]
46. Rouvroye, M.D.; Wicaksono, A.; Bosch, S.; Savelkoul, E.; Covington, J.A.; Beaumont, H.; Mulder, C.J.; Bouma, G.; de Meij, T.G.J.; de Boer, N.K.H. Faecal Scent as a Novel Non-Invasive Biomarker to Discriminate between Coeliac Disease and Refractory Coeliac Disease: A Proof of Principle Study. *Biosensors* **2019**, *9*, 69. [CrossRef] [PubMed]
47. Valli, E.; Panni, F.; Casadei, E.; Barbieri, S.; Cevoli, C.; Bendini, A.; García-González, D.L.; Gallina Toschi, T. An HS-GC-IMS Method for the Quality Classification of Virgin Olive Oils as Screening Support for the Panel Test. *Foods* **2020**, *9*, 657. [CrossRef] [PubMed]
48. Wang, S.; Chen, H.; Sun, B. Recent progress in food flavor analysis using gas chromatography–ion mobility spectrometry (GC–IMS). *Food Chem.* **2020**, *315*, 126158. [CrossRef]

49. Zhang, X.; Dai, Z.; Fan, X.; Liu, M.; Ma, J.; Shang, W.; Liu, J.; Strappe, P.; Blanchard, C.; Zhou, Z. A study on volatile metabolites screening by HS-SPME-GC-MS and HS-GC-IMS for discrimination and characterization of white and yellowed rice. *Cereal Chem.* **2020**, *97*, 496–504. [CrossRef]
50. Haley, L.V.; Romeskie, J.M. GC-IMS: A technology for many applications. *Proc. SPIE* **1998**, *3575*, 375–383. [CrossRef]
51. Kanu, A.B.; Hill, H.H. Ion mobility spectrometry detection for gas chromatography. *J. Chromatogr. A* **2008**, *1177*, 12–27. [CrossRef]
52. Denawaka, C.J.; Fowlis, I.A.; Dean, J.R. Evaluation and application of static headspace–multicapillary column-gas chromatography–ion mobility spectrometry for complex sample analysis. *J. Chromatogr. A* **2014**, *1338*, 136–148. [CrossRef] [PubMed]
53. Ferrer, I.; Thurman, E.M. Liquid chromatography/time-of-flight/mass spectrometry (LC/TOF/MS) for the analysis of emerging contaminants. *TrAC Trends Anal. Chem.* **2003**, *22*, 750–756. [CrossRef]
54. Weber, C.M.; Cauchi, M.; Patel, M.; Bessant, C.; Turner, C.; Britton, L.E.; Willis, C.M. Evaluation of a gas sensor array and pattern recognition for the identification of bladder cancer from urine headspace. *Analyst* **2011**, *136*, 359–364. [CrossRef]
55. Khalid, T.; White, P.; De Lacy Costello, B.; Persad, R.; Ewen, R.; Johnson, E.; Probert, C.S.; Ratcliffe, N. A Pilot Study Combining a GC-Sensor Device with a Statistical Model for the Identification of Bladder Cancer from Urine Headspace. *PLoS ONE* **2013**, *8*, e69602. [CrossRef]
56. Gao, Q.; Lee, W.-Y. Urinary metabolites for urological cancer detection: A review on the application of volatile organic compounds for cancers. *Am. J. Clin. Exp. Urol.* **2019**, *7*, 232–248.
57. Lima, A.R.; Pinto, J.; Carvalho-Maia, C.; Jerónimo, C.; Henrique, R.; Bastos, M.D.; Carvalho, M.; Guedes de Pinho, P. A Panel of Urinary Volatile Biomarkers for Differential Diagnosis of Prostate Cancer from Other Urological Cancers. *Cancers* **2020**, *12*, 2017. [CrossRef]
58. Kurata, Y.; Asamoto, M.; Hagiwara, A.; Masui, T.; Fukushima, S. Promoting effects of various agents in rat urinary bladder carcinogenesis initiated by N-butyl-N-(4-hydroxybutyl)nitrosamine. *Cancer Lett.* **1986**, *32*, 125–135. [CrossRef]
59. Boyes, W.K.; Bingham, E.; Cohrssen, B.; Powell, C.H. Neurotoxicology and behavior. In *Patty's Toxicology*; Wiley: Hoboken, NJ, USA, 2001.
60. Heptanal. Available online: https://pubchem.ncbi.nlm.nih.gov/compound/Heptanal (accessed on 16 October 2021).
61. Wishart, D.S.; Tzur, D.; Knox, C.; Eisner, R.; Guo, A.C.; Young, N.; Cheng, D.; Jewell, K.; Arndt, D.; Sawhney, S.; et al. HMDB: The Human Metabolome Database. *Nucleic Acids Res.* **2007**, *35*, D521–D526. [CrossRef] [PubMed]
62. National Center for Biotechnology Information. PubChem Compound Summary for CID 15979, Pristane. Available online: https://pubchem.ncbi.nlm.nih.gov/compound/Pristane (accessed on 16 October 2021).
63. Rodrigues, D.; Pinto, J.; Araújo, A.M.; Monteiro-Reis, S.; Jerónimo, C.; Henrique, R.; de Lourdes Bastos, M.; de Pinho, P.G.; Carvalho, M. Volatile metabolomic signature of bladder cancer cell lines based on gas chromatography–mass spectrometry. *Metabolomics* **2018**, *14*, 62. [CrossRef] [PubMed]
64. Peng, G.; Hakim, M.; Broza, Y.Y.; Billan, S.; Abdah-Bortnyak, R.; Kuten, A.; Tisch, U.; Haick, H. Detection of lung, breast, colorectal, and prostate cancers from exhaled breath using a single array of nanosensors. *Br. J. Cancer* **2010**, *103*, 542–551. [CrossRef]
65. Chung, W.-G.; Yu, I.-J.; Park, C.-S.; Lee, K.-H.; Roh, H.-K.; Cha, Y.-N. Decreased formation of ethoxyacetic acid from ethylene glycol monoethyl ether and reduced atrophy of testes in male rats upon combined administration with toluene and xylene. *Toxicol. Lett.* **1999**, *104*, 143–150. [CrossRef]
66. Yu, I.-J.; Lee, J.-Y.; Chung, Y.-H.; Kim, K.-J.; Han, J.-H.; Cha, G.-Y.; Chung, W.-G.; Cha, Y.-N.; Park, J.-D.; Lee, Y.-M.; et al. Co-administration of toluene and xylene antagonized the testicular toxicity but not the hematopoietic toxicity caused by ethylene glycol monoethyl ether in Sprague–Dawley rats. *Toxicol. Lett.* **1999**, *109*, 11–20. [CrossRef]
67. Smith, S.; White, P.; Redding, J.; Ratcliffe, N.M.; Probert, C.S.J. Application of Similarity Coefficients to Predict Disease Using Volatile Organic Compounds. *IEEE Sens. J.* **2010**, *10*, 92–96. [CrossRef]
68. Jiménez-Pacheco, A.; Salinero-Bachiller, M.; Iribar, M.C.; López-Luque, A.; Miján-Ortiz, J.L.; Peinado, J.M. Furan and p-xylene as candidate biomarkers for prostate cancer. *Urol. Oncol. Semin. Orig. Investig.* **2018**, *36*, 243.e221–243.e227. [CrossRef] [PubMed]
69. Wu, H.; Liu, T.; Ma, C.; Xue, R.; Deng, C.; Zeng, H.; Shen, X. GC/MS-based metabolomic approach to validate the role of urinary sarcosine and target biomarkers for human prostate cancer by microwave-assisted derivatization. *Anal. Bioanal. Chem.* **2011**, *401*, 635–646. [CrossRef] [PubMed]
70. Lima, A.R.; Pinto, J.; Azevedo, A.I.; Barros-Silva, D.; Jerónimo, C.; Henrique, R.; de Lourdes Bastos, M.; Guedes de Pinho, P.; Carvalho, M. Identification of a biomarker panel for improvement of prostate cancer diagnosis by volatile metabolic profiling of urine. *Br. J. Cancer* **2019**, *121*, 857–868. [CrossRef]
71. Phillips, M.; Gleeson, K.; Hughes, J.M.B.; Greenberg, J.; Cataneo, R.N.; Baker, L.; McVay, W.P. Volatile organic compounds in breath as markers of lung cancer: A cross-sectional study. *Lancet* **1999**, *353*, 1930–1933. [CrossRef]
72. Cai, X.; Chen, L.; Kang, T.; Tang, Y.; Lim, T.; Xu, M.; Hui, H. A Prediction Model with a Combination of Variables for Diagnosis of Lung Cancer. *Med. Sci. Monit. Int. Med J. Exp. Clin. Res.* **2017**, *23*, 5620–5629. [CrossRef]
73. Belugina, R.; Karpushchenko, E.; Sleptsov, A.; Protoshchak, V.; Legin, A.; Kirsanov, D. Developing non-invasive bladder cancer screening methodology through potentiometric multisensor urine analysis. *Talanta* **2021**, *234*, 122696. [CrossRef] [PubMed]
74. Kim, Y.-H.; Kim, K.-H. A novel method to quantify the emission and conversion of VOCs in the smoking of electronic cigarettes. *Sci. Rep.* **2015**, *5*, 16383. [CrossRef] [PubMed]

75. Charles, S.M.; Batterman, S.; Jia, C. Composition and emissions of VOCs in main-and side-stream smoke of research cigarettes. *Atmos. Environ.* **2007**, *41*, 5371–5384. [CrossRef]
76. McWilliams, A.; Beigi, P.; Srinidhi, A.; Lam, S.; MacAulay, C.E. Sex and Smoking Status Effects on the Early Detection of Early Lung Cancer in High-Risk Smokers Using an Electronic Nose. *IEEE Trans. Biomed. Eng.* **2015**, *62*, 2044–2054. [CrossRef] [PubMed]

 biosensors

MDPI

Review

On Demand Biosensors for Early Diagnosis of Cancer and Immune Checkpoints Blockade Therapy Monitoring from Liquid Biopsy

Sai Mummareddy [1,†], Stuti Pradhan [2,†], Ashwin Kumar Narasimhan [3] and Arutselvan Natarajan [4,*]

1 Department of Biology and Chemistry, Emory University, Atlanta, GA 30322, USA; sai.sreehith.mummareddy@emory.edu
2 Department of Microbiology, Immunology, and Molecular Genetics, University of California, Los Angeles, CA 90095, USA; stutipradhan@ucla.edu
3 Department of Biomedical Engineering, SRM Institute of Science and Technology, Chennai 603203, India; ashwinkn05@gmail.com
4 Molecular Imaging Program at Stanford (MIPS), Department of Radiology, Stanford University, Stanford, CA 94305, USA
* Correspondence: anatarajan@stanford.edu; Tel.: +1-650-736-9822
† These authors contributed equally to this work.

Citation: Mummareddy, S.; Pradhan, S.; Narasimhan, A.K.; Natarajan, A. On Demand Biosensors for Early Diagnosis of Cancer and Immune Checkpoints Blockade Therapy Monitoring from Liquid Biopsy. *Biosensors* **2021**, *11*, 500. https://doi.org/10.3390/bios11120500

Received: 11 October 2021
Accepted: 1 December 2021
Published: 7 December 2021

Publisher's Note: MDPI stays neutral with regard to jurisdictional claims in published maps and institutional affiliations.

Abstract: Recently, considerable interest has emerged in the development of biosensors to detect biomarkers and immune checkpoints to identify and measure cancer through liquid biopsies. The detection of cancer biomarkers from a small volume of blood is relatively fast compared to the gold standard of tissue biopsies. Traditional immuno-histochemistry (IHC) requires tissue samples obtained using invasive procedures and specific expertise as well as sophisticated instruments. Furthermore, the turnaround for IHC assays is usually several days. To overcome these challenges, on-demand biosensor-based assays were developed to provide more immediate prognostic information for clinicians. Novel rapid, highly precise, and sensitive approaches have been under investigation using physical and biochemical methods to sense biomarkers. Additionally, interest in understanding immune checkpoints has facilitated the rapid detection of cancer prognosis from liquid biopsies. Typically, these devices combine various classes of detectors with digital outputs for the measurement of soluble cancer or immune checkpoint (IC) markers from liquid biopsy samples. These sensor devices have two key advantages: (a) a small volume of blood drawn from the patient is sufficient for analysis, and (b) it could aid physicians in quickly selecting and deciding the appropriate therapy regime for the patients (e.g., immune checkpoint blockade (ICB) therapy). In this review, we will provide updates on potential cancer markers, various biosensors in cancer diagnosis, and the corresponding limits of detection, while focusing on biosensor development for IC marker detection.

Keywords: cancer markers; immune checkpoints; PD-1; PD-L1

1. Introduction

Cancer is one of the leading causes of death worldwide. The high incidence of cancer and the corresponding elevated mortality rate has made the creation of new diagnostic tools and therapeutic techniques a high clinical priority. Prognosis is best when cancers are detected early, but this task is challenging due to the difficulties of detecting small, early-stage tumors [1]. Tumor-specific signature markers expressed on the cell surface, i.e., cancer biomarkers, can help in diagnosing cancer early. Combining these cancer marker signals with biosensing tools (known as biosensors) should enable early cancer detection [2].

As a device, biosensors integrate analytes, receptors, transducers, and outputs to measure the expression level of the cancer markers. These analytical and diagnostic devices can be used to process biological samples (analytes) to collect specific information using a combination of biological detecting molecules (receptors) and an electronic sensor system with a transducer (Figure 1). Biosensors are developed based on the target analyte, such as cancer markers, immune markers, and DNA, found in biological samples. Some examples of biosensors include bio-computers, glucometers, biochips, and resonant mirrors [3].

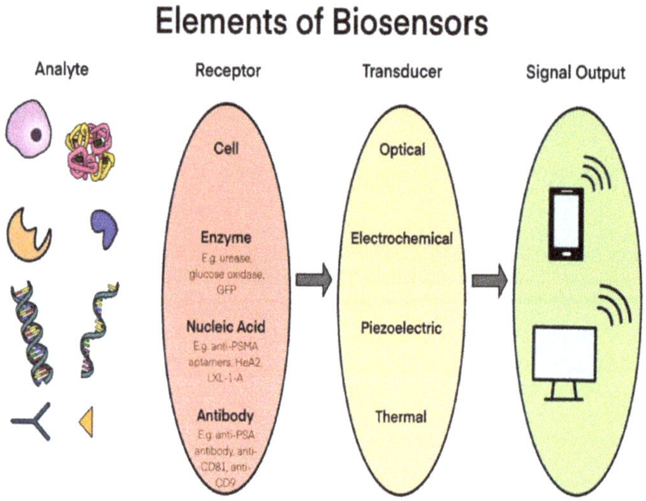

Figure 1. General schema of biosensor device workflow.

In this schema we have depicted the significant parts in sequence used in for any biosensor device. It shows analytes from biological samples input to signal output from digital displays by providing a few examples of analyte molecules such as proteins, DNA and other targeting agents; receptors expressed on cells, enzymes, nucleic acid, and antibodies [4]. In addition, this illustration indicating the well-known transducers used in biosensors, i.e., optical, electrochemical, piezoelectric, and thermal. Specific information on each parts of biosensor is provided in the corresponding sections of this review.

Biosensors are designed to detect specific biological markers or analytes (i.e., proteins, DNA, RNA and cells) and convert biological molecule interaction-signals into an electrical signal that can be measured as a digital output. In addition, biosensor technology has the potential to deliver fast and accurate information, as well as measure cancer cells and cancer metastasis. It can also be used to determine the therapeutic effectiveness of anticancer drugs, assess cancer biomarkers, and determine effectiveness of drugs at various target sites. Biosensors are an emergent tool for various disease management with great potential in cancer detection and monitoring. Overall, these biosensors are made to reduce the diagnostic time for a patient's disease and to monitor therapeutic outcomes.

There are several reviews available elsewhere with respect to the biosensor device instrumentation, technology, and engineering for signal processing [5–7]. Hence, in this review, we would like to illuminate two key areas related to biological components: (a) the various biomarkers used as analytes to measure the disease conditions and (b) types of sensors and transducers that could detect biomarkers at the lowest level. The focus is to further the understanding of ICB and their potential use in the biosensor development. These ICBs play a major role in understanding the early diagnosis and treatment efficacy post chemo or radiotherapy. A detailed analysis of check point markers is discussed for use in future biosensor development.

2. Key Methods for the Detection of Cancer Biomarkers from Liquid Biopsy

Detection of biomarkers and immune checkpoints in liquid biopsies can be analyzed with various transduction principles, targets, and analytes, as summarized in Figure 1. Analytes are designed to bind to a specific receptor, protein or biomarkers on cells. Binding between the target and analyte depends on the transduction principle and the sensor type which produces the corresponding output. Finally, the output signal is processed and analyzed to display on the device. Several techniques are widely used for the detection of various disease markers, such as optical, electrochemical, piezoelectric, and thermal based sensors. Among these detection methods, optical and electrochemical based biosensors are cost effective and highly sensitive with low detection limits and high reproducibility. The basic principle of optical biosensors works on the interaction between antigens and antibodies, where the binding affinity intensity is transformed into proportional electronic signals detected in the transducer unit. The optical sensing unit consists of a laser source, a spectrometer, an immobilized sensor, and an electronic device to amplify the interaction [8]. Optical sensors work based on two methods: (1) Direct detection and (2) Indirect detection through exogenous agents. In both methods the primary signal measurements are derived from changes in the absorption and fluorescence intensity, colorimetric, mechanical sensors such as microcantilevers and variations in refractive index [9]. Furthermore, these detection systems can be applied by combining the opto-electronic device and lab on chip technologies.

Similarly, electrochemical-based detectors convert chemical energy into electric potentials. To measure this chemical energy, electrodes are used as a transducer, i.e., electrodes are coated with receptors that interact with analytes. When a redox reaction occurs between the analyte and the receptor, the external voltage is applied to the transducer element (electrode), which generates a current. This current is further amplified via signal process tools to identify the desired chemical reaction [10]. Electrochemical sensors are classified based on the detection output method, such as amperometric, potentiometric and voltametric, as well as on enzymatic and non-enzymatic-based detection in liquid biopsies [10,11]. In the next section, we will discuss the various proteins and antigens used for the detection of tumors.

Detecting cancer using biomarkers from circulating fluids such as the blood has received remarkable attention in recent years. Liquid biopsy sampling is minimally invasive and requires a small volume of blood to detect and characterize tumors and monitor treatment outcome [12]. A broad variety of cancer markers and associated assays have been under development to diagnose cancer via different detecting systems. For example, various blood-based biomarkers such as tumor associated antigens (TAAs, Figure 2), circulating tumour cells (CTCs), and circulating cell-free tumor DNA (ctDNA) are widely utilized.

Biosensors are designed to detect specific biological markers or analytes (i.e., proteins, DNA, RNA and cells) and convert biological molecule interaction-signals into an electrical signal that can be measured as digital output. In addition, biosensor technology has the potential to deliver fast and accurate information, as well as measure cancer cells and cancer metastasis. It can also be used to determine the therapeutic effectiveness of anticancer drugs, assess cancer biomarkers, and determine effectiveness of drugs at various target sites. Biosensors are an emergent tool for various disease management with great potential in cancer detection and monitoring. Overall, these biosensors are made to reduce the diagnostic time for a patient's disease and to monitor therapeutic outcomes. There are several reviews available elsewhere with respect to the biosensor device instrumentation, technology, and engineering for signal processing [5–7]. Hence, in this review, we would like to illuminate two key areas related to biological components: (a) the various biomarkers used as analytes to measure the disease conditions and (b) the types of sensors and transducers that could detect biomarkers at the lowest level. The focus is to further the understanding of ICB and their potential use in the biosensor development. These ICBs play a major role in understanding the early diagnosis and treatment efficacy post chemo or radiotherapy. A detailed analysis of check point markers is discussed for use in future biosensor development.

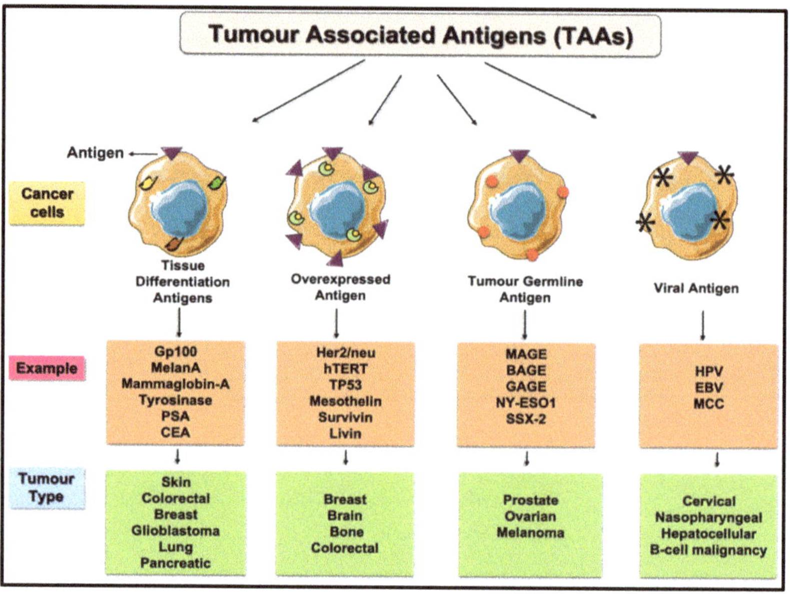

Figure 2. Tumor associated antigens and its expression pattern.

2.1. Tumor-Associated Antigens (TAAs)

TAAs can be divided into three primary categories: (1) normal proteins that are overexpressed by cancer cells, (2) differentiation antigens, and (3) cancer testis antigens (CTA) (Table 1) [13]. TAAs (Figure 2) and tumor-specific antigens (TSAs) have been studied for decades due to their presence during early stages of cancer. While TSAs are found solely on cancer cells, TAAs are also found on healthy cells but are overexpressed on cancer cells, allowing them to be used to detect cancer. In most instances, the biological system may produce antibodies to combat the TAAs, thereby providing additional opportunities for diagnostic markers. In 1943, Gross et al., conducted an experiment determining the ability of acquired immunity to combat malignant tumors by detecting TAAs [14]. Since then, various TAAs have emerged as potentially effective biomarkers for early detection of cancer. However, rigorous validation protocols are required to make these potential TAAs into early cancer biomarkers.

An ideal TAA should be specific to a type of cancer—it should not be measurable in control or normal tissues, in patients with other tumors, or in those with autoimmune diseases. However, many TAAs lack this specificity. For example, p53 is a tumor suppressor mutated in cancer patients, but it is also detectable in various tumors and in autoimmune diseases [15]. Therefore, rigorous biomarker validation is necessary [16]. Determining the immunogenicity of TAAs allows antigens with the greatest potential for diagnostic and therapeutic purposes to be separated from those with low specificity and sensitivity. When detecting TAAs using biosensing devices, it is important to choose biomarkers with attributes such as high sensitivity, specificity, stability, and reproducibility in the biological system. Furthermore, the signal corresponding to the marker should provide insights into the origin, stage, and progression of disease. For example, Silva and his coworkers have attempted the detection of TAAs such as peanut agglutinin in serum samples at 0.01 mg mL^{-1}, which is a comparatively lower limit of detection (LOD) than other sensors [17]. Thus, the determination of TAAs using biosensor devices at low LOD is still in its infant stage.

Table 1. Candidates for Early Detection Biomarkers to Cancer Phenotype: Advantages and Limitations.

Types of Antigens	Antigen	Cancer Type	Strengths for Early Detection	Limitations for Early Detection
Overexpressed Antigen	MUC1	Breast	Expressed in over 90% of breast tumors	Most prevalent in metastasized breast cancers
	HER2	Breast/Esophageal Gastric/Ovarian Endometrial	Nearly 2 million receptors expressed on tumor cell surface	Can cause toxicity to healthy cells given low expression of antigen
Differentiation Antigen	gp100	Melanoma	High expression in malignant glioma cells	Expression in normal brain tissues
	CEA	Colorectal cancer	Greater sensitivity than other diagnostic methods	Limited sensitivity Most effective in detection of recurrent cancers
CT Antigen	MAGE-A	Melanoma Squamous cell carcinomas	Very specific to cancer cells	Higher prevalence in metastatic cancers

2.2. Overexpressing Cancer Biomarkers

Several TAA biomarkers are overexpressed in cancer cells and a few examples are described as follows. Mucin-1 (MUC1) is a protein biomarker overexpressed in 90% of breast tumors. The N-terminal of MUC1 is shed from carcinoma cells and can be found in the plasma of women with breast cancer [18]. However, MUC1-N prevalence is greatest in breast cancers that have metastasized, so it is not an ideal target for early detection. On the other hand, human epidermal growth factor receptor-2 (HER2), a well-studied protein overexpressed in cancer cells are potential for targeted therapy [19]. Overexpression of HER2 has been found to directly correlate to poor prognosis of numerous cancers, including breast, esophageal, gastric, ovarian, and endometrial cancer [20]. HER2 overexpression is found in 15–30% of breast cancers which can have up to 25 to 50 copies of the HER2 gene, resulting in nearly 2 million receptors on the tumor cell surface [21]. The extracellular domain of the HER2 protein is cleaved and can be detected in blood serum using ELISA. However, when these antigens have relatively low expression, targeting them during treatment can cause toxicity in healthy tissues. For example, the number of receptor copies expressed on breast cancer cells can be classified as HER2 positive, negative or equivocal, based on the IHC results [21].

2.3. Differentiation Antigens

Differentiation antigens are expressed by cells during specific stages of cell development and in tissue samples. These differentiation antigens are considered as a new type of cancer biomarker for biosensing. Their specificity indicates a strong potential for differentiation antigens to be detected for accurate cancer diagnosis. For example, glycoprotein-100 (gp100) is a differentiation antigen presented on melanocytic cells in large amounts that can be targeted by anti-melanoma cytotoxic T lymphocytes (CTLs). Additionally, gp100 is immunogenic, and it can induce an immune response, suggesting that melanocyte differentiation antigens may be tumor rejection antigens [22].

Carcinoembryonic antigen (CEA) is another differentiation antigen widely used for cancer screening protocols as an elevated level of serum CEA corresponds to colorectal cancer (CRC) in approximately 17–47% of individuals [23]. It can be used as an independent prognostic factor and can predict outcomes of patients with stage II CRC. CEA has also shown greater sensitivity for detecting recurrence in comparison to other diagnostic techniques such as CT scans. In a study performed by Tsikitis et al., postoperative surveillance of CEA detected first recurrences in 29.1% of patients with early-stage colorectal cancer, while CT scan detected 23.6%, colonoscopy detected 12.7%, and chest X-rays detected 7.3% [24]. The sensitivity of CEA makes it an ideal biomarker to detect colorectal cancer. It is the best-known marker to detect tumor recurrence. Monitoring CEA levels in patients who have recovered from CRC could be an effective screening method (for recurrence). Nevertheless, while CEA has the greatest sensitivity when compared to other diagnostic techniques, it is still unable to detect most CRC recurrences, indicating a need for better tumor-associated antigens to be discovered. Several biosensors have been developed to increase the efficiency of the detection of CEA using nanoparticles. Using anti-CEA as targeted moieties with different sensing mechanism like electrochemical, optical, chemiluminescence, and acoustic wave biosensors were studied [25]. However, there are still restrictions in the detection limit at pico to nano gram level from the samples.

2.4. Cancer Testis Antigens

Cancer testis antigens (CTA), classified under tumor germline antigens, are found in the male germ cells of healthy adults. However, CT antigens are also found in tumor cells of various types of cancer, classifying them as neoantigens, or a protein that forms on cancer cells previously unknown to the body. CTA antigens lack MHC class 1 molecules. This allows them to be immune privileged, meaning that they can tolerate the introduction of antigens without causing an immune response. However, the cellular and humoral immune responses observed in cancer patients displaying CTA antigens as well as the correlation of CTA antigens with cytolytic activity of tumor immune infiltrates suggests that CTA antigens are highly specific immune targets [26]. Melanoma-associated antigen-A (MAGE-A) is a CT antigen expressed in 32% of melanomas and 45% of squamous cell carcinomas. The high expression of MAGE-A indicates that it may be a potential target for detection and immunotherapy [27]. However, it was observed that MAGE-A is more prevalent in metastatic cancers than primary tumors, indicating its lack of efficiency towards early detection.

2.5. Prostate Specific Antigen (PSA) Based Biosensor

Prostate specific antigen (PSA) is the most common cancer marker for the detection of prostate cancer from serum. The PSA biosensor is primarily developed with a combination of the PSA targeting receptors and a transducer (e.g., electrochemical, micro-cantilever, and surface plasmon resonance) to detect prostate cancer.

The changes in vibrational frequency upon the interaction of an antigen binding to an antibody are used to detect PSA [28,29]. Several approaches have been investigated to build biosensors to diagnose prostate cancer non-invasively with greater sensitivity than the standard ELISA technique [30].

Integration of conventional biosensors with nanomaterials increase the sensitivity and specificity of such markers. For instance, the use of gold electrodes, graphene or graphene-oxide, carbon nanotubes, hybrid nanoparticles, and other types of nanoparticles are potential agents for PSA detection. The most attractive sensing schemes are summarized in a Table 2.

Table 2. List of Different Biosensors, their Advantages and Limitations and their Potential Biomarker Targets.

Biosensor Types	Advantages	Disadvantages	Examples of Biomarkers Detected	References
Localized and Compact Surface Plasmon Resonance Biosensor (LSPR and CSPR)	Highly sensitive and specific Label-free system Real-time measurement and detection Usage of gold nanomaterials Small sample size required to run	Complex mechanism of action Cost-effectiveness Non-target binding Immobilization on surface causing configurational changes	PD-L1 MT1-MMP IFN-γ PSA IgG TNF-α CRP	[31–34]

Table 2. *Cont.*

Biosensor Types	Advantages	Disadvantages	Examples of Biomarkers Detected	References
Electrochemical Biosensor	Cost and time- effective Small sample size required to run Very easy to use system for consumers Really good detection limits	Reproducibility issues (either one or few time usage) Not as sensitive as the other biosensors and conventional methods Very low shelf life	CEA NSE MUC1 EpCAM Multiple Types of miRNAs BRCA1 HER2	[35–38]
Colorimetric Biosensor	Cost and time-effective Very easy-to use Small sample size required Portable and easy to maintain	Low sensitivity and specificity Reproducibility issues Low shelf life Limited multiplexing and quantification capabilities	CEA AFP PSA miRNA-148a miR-21 miR-155	[39–41]
Multiplexed Nanobiosensors	Cost-effective High sensitivity and specificity compared to conventional methods Parallel detection of checkpoint markers (gives more insight in checkpoint interactions) Very small sample size needed	Still need to find effective ways for the target to get to the surface rapidly Non-target binding on the surface Non-specific interferences	PD-1 PD-L1 LAG-3 miR-21 miR-574-3p EpCAM Bladder Cancer MicroRNAs	[42–44]

Among approaches that results in the highest sensitivity is the electrochemical-based biosensor (as it can detect PSA at attomolar concentrations), followed by mass cantilever sensing and the electro-chemiluminescent (EC) approach. EC-based biosensors can be modified by changing the antibodies and by using other binders such as aptamers, peptides, and nanobodies with the same device [45]. For example, horseradish peroxidases (HRP), alkaline phosphatase (ALP), glucoamylase, β-galactosidase, or glucose oxidase are conjugated to antibodies (Ab) and can be employed. In this approach, the $Ab_{[secondary]}$ is directly tagged to the reporter enzymes, which catalyze the reduction of their respective substrates in the presence or absence of a mediator [46]. The HRP enzyme-linked electrochemical sandwich assay is a well-known and widely used assay due to its stability and fast reaction kinetics, which catalyzes the reduction of H_2O_2 in the presence of hydroquinone (HQ) [47]. However, it produces a background signal in the absence of the analyte due to endogenous electrochemical activity of its H_2O_2 substrate. Additionally, HRP relies on the use of mediators for signal generation. ALP has been used to overcome the background and mediator limitations of HRP.

The EC-based indirect detection of PSA assay was developed with incorporation of multiple receptor elements, NPs, redox labels, or other biomolecular processes. These PSA biosensors provided a basis for multi-target analysis with increased sensitivity and selectivity. For example, the detection limit demonstrated clinically relevant LODs ranging from 1 ng mL^{-1} to 0.020 fg mL^{-1} (Table 2). Using these devices, a linear range of 0.001–10 ng mL^{-1} and a limit-of-detection (LOD) of 0.84 pg mL^{-1} were obtained for detecting PSA in serum. In related study, Chen et al. developed gold screen-printed electrode-based microfluidic devices (GSPE-MFDs) to demonstrate the detection and quantification of PSA in human serum samples [48]. In this system, GSPE Ab_c conjugated magnetic particles (MNPs) were immobilized for capturing PSA, followed by HRP-Ab_s captured on the PSA-attached electrode using the microfluidic device. The enzymatic and electrochemical reaction between HRP and HQ generates current signals for measurement. Similarly, Zani et al. utilized a magnetic bead-based assay by using mouse IgG-Ab_c for PSA capture and ALP tagged anti-mouse IgG for signal transduction [49]. This PSA-targeted biosensor was able to detect as low as 2 ng mL^{-1} in human serum samples. Furthermore, it shows a linear range of PSA detection (1–80 ng mL^{-1}). Overall, enzyme-based sandwich assays demonstrated very high specificity to PSA, but they also have intrinsic enzymatic instability issues [50]. In addition, these assays require several washing steps to remove non-specific binding to avoid false negative or false positive results.

3. Immune Checkpoints as a Marker for Cancer Diagnosis

Immune checkpoints and their use in cancer immunotherapies is a rapidly growing field of study [51]. Immune checkpoint-based treatment increases the hope among the cancer patients and especially those who did not respond to more established treatments [51]. Currently, the surface markers, programmed death-1 (PD-1) and CTLA-4 are well known for their function in the immune system, as well as their role in cancer as theranostic applications [52]. PD-1 and CTLA-4 are two major checkpoints approved by the FDA for immune checkpoint blockade therapy [53]. Similarly, other checkpoints, including TIM-3, BTLA, and LAG-3, are under active investigation as potential biomarkers for cancer theranostics [54]. These immune checkpoints can be detected on tumor cells and in the tumor microenvironment (TME) through traditional tissue biopsies and IHC as well as via biosensors. Although these biomarkers are potentially for the detection of cancer, the developmental status of biosensors designed to detect these immune checkpoint markers is still at an early stage [55]. In this article, we will focus upon several such soluble immune checkpoints-sPD-1, sPD-L1, sLAG-3, and sTIM-3-that are expressed on peripheral blood mononuclear cells that could potentially be diagnosed through biosensors.

3.1. T-Cell Immunoglobulin and Mucin Domain-3 (TIM-3)

TIM-3 is part of the TIM family and is a receptor found on interferon-γ-producing CD4+ and CD8+ T cells [56]. Initially, TIM-3 was considered primarily as a co-inhibitory checkpoint; a protein that is co-regulated and co-expressed alongside PD-1, LAG-3, and TIGIT [57]. Later, TIM-3 emerged as a co-stimulatory checkpoint depending on the cell type on which it is expressed and the immune response. It is also involved in T-cell exhaustion and therefore found in tumors [58]. In a study of leukocytes from peripheral blood mononuclear cells, the level of TIM-3 expression was significantly higher in the CD4+ and CD8+ T cells of ovarian cancer patients. CD4+ T-cells displayed further elevation of TIM-3 in ovarian cancer recurrence, suggesting TIM-3 as a biomarker for early detection as well as detection of recurrence [59].

TIM-3 upregulation in CD4+ and CD8+ peripheral T cells was found to correlate with the presence of osteosarcoma. The level of TIM-3 expression also increased with tumor stage and metastasis, and higher TIM-3 levels were associated with worse overall survival. Such findings suggest TIM-3′s value as a diagnostic and prognostic marker for osteosarcoma [60]. In hepatocellular carcinoma, TIM-3 expression is induced by other cytokines located in proximity in the tumor microenvironment (TME). However, TIM-3 expression, along with the expression of other T cell exhaustion markers such as PD-1 and CTLA-4, is lower in the non-tumor microenvironment (NTME) and the peripheral blood than the TME [61]. The variation in expression of TIM-3 in the TME and NTME presents a challenge for using immune checkpoints to detect cancer non-invasively.

3.2. B- and T-Lymphocyte Attenuator (BTLA)

BTLA is a lymphocyte inhibitory receptor found on B-, T-, and other mature lymphocyte cells. BTLA regulates the immune system by inhibiting T cell reactions and restricting cytokine production and creates a tumor microenvironment that suppresses immune responses [62]. Reports indicating high BTLA expression were found in gastric cancer, pancreatic cancer, and chronic lymphocytic leukemia (CLL)/small lymphocytic leukemia [63]. However, this upregulation does not correlate to the presence of BTLA in the bloodstream. For example, a higher gene transcript level is observed in CLL patients. BTLAs were detected in low level peripheral blood B cells and normal BTLA levels were detected on T cells. Even in-vitro stimulation resulted in lower levels of BTLA expression on B and T lymphocytes [64].

Different stages of B and T cells developed from bone marrow, which include pro-B/T, pre-B/T and immature B/T cells, each express unique marker. T cells are activated by B cells and further divided into T-helper and T-regulatory cells. In 1994, the number of B-cells were determined using piezoelectric sensors by immobilizing anti-B cells on the surface electrode [65]. However, capturing B-cells at high accuracy requires sensitive biosensors to obtain an amplified signal. T-cell lymphocytes CD4+ detection using nano based biosensors was explored recently by Carinelli et al. using magnetic particles [66]. Here, isolation of CD4+ cells were captured magnetically via anti-CD4+ using magnetic particles [66]. Similarly, single wall carbon nanotubes-based detection coupled with electrochemical biosensor for the detection of CD4+ using with a limit of detection at 1×10^2 cells mL^{-1} [67]. Chen and his coworkers developed a rapid assay biosensor to detect dynamic T-cell activation using nanoparticles, wherein, polyaniline fibers coated with gold nanoparticles detected multiple immunosensing markers such as CD69, CD25, and CD71 with specific anti-CD molecules. Time dependent activation of immune markers were captured with an LOD of 1×10^4 cells mL^{-1} [68]. Hence, understanding the role of T-cell immuno-sensors at high LOD is important to reveal the immune inhibition of activated drugs.

3.3. Immune Markers in Serum

Immune markers serve as potential prognostic biomarkers in serum and could have significant predictive power of disease progression in personalized medicine. While some immune checkpoints can be detected on peripheral blood cells (e.g., PD-1) and others (PD-1, PD-L1, LAG-3, and TIM-3) can be detected in serum due to ectodomain cleavage. This section reviews such potential immune markers for the serum.

3.4. Lymphocyte-Activation Gene 3 (LAG-3)

A disintegrin and metalloproteinase domain-containing protein (ADAM) such as ADAM-10 and ADAM-17 cleave the extracellular domain of LAG-3 [69]. Higher levels of serum LAG-3 (sLAG-3) were found to correlate with earlier stages of non-small cell lung cancer. Serum of patients with stage III and IV cancer had lower levels of sLAG-3 comparatively [70]. Contrastingly, when serum sLAG-3 levels were assessed in gastric cancer patients, individuals with gastric cancer were found to have sLAG-3 levels of 247.52 ± 51.28 ng/mL while healthy individuals had sLAG-3 levels of 869.46 ± 64.35 ng/mL. Additionally, sLAG-3 was found to have a greater sensitivity and accuracy in comparison to other serum biomarkers for gastric cancer such as CEA. sLAG-3 had a sensitivity of 88.60% and an accuracy of 90.65% when evaluated at a threshold value of 378.330 ng/mL. On the other hand, CEA had a sensitivity of 56.20% and an accuracy of 68.61% at a threshold value of 21.755 ng/mL [71]. Measures of soluble LAG-3 in serum have prognostic value, for example, high levels of soluble LAG-3 in some subsets of breast cancer correlate with disease-free, metastasis-free, and overall survival [72].

3.5. PD-1

Serum PD-1 or sPD-1 (along with sPD-L1) is elevated in the serum of melanoma patients in comparison to healthy blood donors [73]. Higher sPD-1 levels are also found in non-small cell lung carcinoma (NSCLC), diffuse large B-cell lymphoma (DLBCL), chronic lymphocytic leukemia (CLL), nasopharyngeal carcinoma (NPC), HCC, pancreatic adenocarcinoma, and advanced rectal cancer [74]. Additionally, sPD-1 and sPD-L1 can be used to predict the outcome of certain immunotherapies, such as anti-PD-1 treatment. A threshold of 500 pg/mL was used to differentiate progression-free survival rates, because of the finding that high baseline serum sPD-1 levels were associated with poor survival [73]. Similarly, serum sPD-1 and sPD-L1 levels were significantly elevated in triple-negative breast cancer patients prior to neoadjuvant chemotherapy. However, patients who responded well to the treatment (as observed through complete or partial remission) had decreased serum sPD-1 and sPD-L1 levels compared to those who had a poor response to the treatment [75].

3.6. PD-L1

The soluble programmed-death ligand 1 (aka PD-L1) is an immune biomarker that can potentially be used for early cancer diagnosis research due to its important role in cancer immunoregulation. The PD-L1 marker essentially binds to its receptor of PD-1 and partakes in pleiotrophic signaling pathways. When PD-L1 binds to the PD-1, there is a halt in the T cells going out to kill the foreign antigens. Thus, studies have shown that the inhibition of this interaction between the PD-L1 marker and the receptor PD-1 can result in drastic reduction of tumor growth and promote antitumor immunity. Thus, it is important to detect the concentrations of the PD-L1 marker in blood. However, it is difficult to quantify the concentration of PD-L1 marker in blood due to signal sensitivity at low concentrations.

Localized surface plasmon resonance (LSPR) biosensors enable one to improve the sensitivity of the signal at low concentrations of PD-L1 using metal nanoparticles [31]. The detection of PD-L1 can be seen through the wavelength shift in the biosensor where the anti-sPD-L1 antibodies get immersed and localized with the gold nanoparticles in the biosensor. In this study, gold nano-shells were developed with LSPR at NIR region around 1600 nm to detect PD-L1. At different serum media, the biosensor showed high specificity to PD-L1 with a sensitivity of 1 pg mL^{-1}. Thus, gold nano shells-based detection of PD-L1 show high sensitivity compared to other conventional techniques like ELISA, IHC and PCR. PD-L1 undergoes proteolytic cleavage by ADAM 10 and ADAM 17 in breast cancer, resulting in a ~37 kDa fragment that can be detected in the media [76,77]. sPD-L1 is elevated in patients with papillary thyroid cancer (PTC) compared to healthy patients. Serum sPD-L1 levels were found to be 0.37 ng/mL in healthy individuals and 0.48 ng/mL in cancer patients. Furthermore, higher sPD-L1 levels were indicative of shorter disease-free survival, suggesting the value of sPD-L1 as a prognostic marker of PTC [78]. Even in primary central nervous system lymphoma, serum sPD-L1 levels (0.429 ng/mL) were higher than those in healthy individuals (0.364 ng/mL). Higher sPD-L1 levels resulted in more frequent relapse (78% of the study participants) than the lower sPD-L1 levels (50%) [79].

ADAM 10 and ADAM 17 are also major target of TIM-3 that result in soluble TIM-3 (sTIM-3) [80]. Serum sTIM-3 is found to be higher in patients with osteosarcoma when compared to patients with benign tumors and the control group (14.4 ng/mL $\pm$ 2.9 vs. 10.3 ng/mL $\pm$ 1.7 vs. 6.3 ng/mL $\pm$ 1.9). Benign tumors also displayed higher sTIM-3 levels than the control, suggesting a possibility for early detection for prevention [81]. Additionally, elevated sTIM-3 levels were also associated with hepatocellular carcinoma (HCC). The control group had a mean serum sTIM-3 level of 2.64 $\pm$ 0.32 pg/mL while patients with HCC had a mean serum sTIM-3 level of 3.57 pg/mL $\pm$ 0.22 [82].

4. Biosensors for Cancer Markers and Immune Checkpoint Detection

Biosensors that have been used in the detection of cancer and immune checkpoint markers include electrochemical and colorimetric biosensors. Most of these biosensors share the following characteristics: consumer friendly, easy to use systems, moderate cost, low turnaround time for the results, and minimal sample size required for testing. However, when compared to detecting markers in samples with low concentrations these biosensors are not as sensitive or specific in comparison to the SPR biosensors [83,84]. Furthermore, these types of sensors can only be used once or a limited number of times, reducing their value for conducting multiple runs.

PD-1 and PD-L1 are known [85,86] to be independent prognostic factors for several tumor associated markers for immunotherapy. Kruger et al. quantified sPD-1 and sPD-L1 in pancreatic cancer serum with ELISA using the human PD-1 antibody duo sets for ELISA development [87]. Reporting on immune check point of therapy with PD-L1 alone is insufficient and showed poor outcome in pancreatic cancer. For example, higher concentrations of PD-L1 and PD-1 were identified in 41 patients to test for advanced stages of carcinoma. Further studies are required to understand the correlation between the two prognostic factors for immune check point blockade therapy. Similarly, a multiplex immunoassay detection (8-color flow cytometry) method was developed using cryopreserved samples of healthy human donors to determine the level of CD8+ and CD4+, for early cancer diagnosis [88]. Upon stimulation of T cell lymphocytes, measurement of upregulated soluble isoforms of LAG-3, TIM-3, CTLA-4, and PD-1, as detected in flow cytometry, directly correspond to treatment response.

4.1. Biosensors for Detection of Multiple Cancer/IC Markers

Multiplexed immune checkpoint biosensor (MICB) development is progressing rapidly due to their high specificity and selectivity [3]. Multiplex biosensors require small volumes of sample to diagnose the disease and work simultaneously to detect multiple immune checkpoints. The parallel detection device is growing as it can be potentially translatable for routine clinical applications. However, with these sensors, the analyte tends to take longer to bind with the cell surface without the use of nanofluidic mixing. Hence, these sensors are time consuming and generate a response slowly. For example, a droplet loaded in the electrode can detect the PD-L1 marker and can run multiple samples at once (28 for the multiplexed biosensor on the PD-L1 marker) [42]. Advanced versions of multiplexed electrohydrodynamic biosensors have shown improved performance for biomarker detection [89,90]. However, many of the multiplexed lateral flow nanosensors display this limitation. Additionally, getting and visualizing the target alone for these sensors is also difficult as there are many non-specific interferences displayed in the system. Based on the applications and sample results, biosensors can be broadly classified into two types: (a) high sensitivity and specificity at the cost of a complex system (SPR and multiplexed biosensors) and (b) easy to use and cost and time effective, but not possessing high sensitivity and specificity (colorimetric and EC biosensors).

Another MICB developed by Wuethrich et al. detects soluble PD-1, PD-L1, and LAG-3 in parallel in liquid biopsies, using as little as a single sample drop of approximately 20 μL volume per target immune checkpoint [42]. MICB utilizes a microfluidic sandwich immunoassay with high-affinity yeast cell-derived single chain variable fragments (ScFvs) as well as alternating current electrohydrodynamic in-situ nanofluidic mixing. The high-affinity yeast cell-derived ScFvs are an alternative to the use of monoclonal antibodies and polyclonal antibodies, which are costly and have a limited shelf life. High-affinity yeast cell-derived ScFvs provide longer shelf life without sacrificing specificity and can be mass-produced for a lower cost. The nanofluidic mixing stimulates immune checkpoint interactions with the biosensor while diminishing non-target sensor binding, providing greater specificity for immune checkpoint capture.

MICB has ability to analyze 28 samples in less than two hours. To understand the ability for MICB to detect soluble PD-1, PD-L1, and LAG-3 in liquid biopsies, various concentrations of immune checkpoints were spiked into diluted human serum. While the assay was able to detect a differentiable signal response from the samples with the immune checkpoints, fortified samples had a concentration equal to the prognostic cut-off, which may differ from the threshold for cancer detection [42].

4.2. Exosome Biosensor

Exosomes are the key players in communication at the intercellular level through surface biomarkers. For instance, exosomal proteins are fundamental in cancer development, drug resistance, and differentiation etc [91]. Expression of PD-L1 in exosomes is a biomarker that is overly expressed in lung tumors. Hence, many exosomal proteins that include exosomal immune markers like PD-L1, can be promising cancer biomarkers. In a recent study, researchers constructed a compact surface plasmon resonance biosensor that detects exosomal PD-L1 and EGFR [92]. Looking at the different serum samples of patients with lung cancer, researchers used this biosensor to determine the specificity and sensitivity of the biomarker in these samples.

This biosensor worked through optical resolution using laser beams. Essentially, two laser beams were used in which one allowed for the SPR to interact with the exosomal proteins on the surface of the sensor. The other beam was used as reference, and the two photodetectors quantified the exosomal proteins by magnifying the intensity of the initial beam and the reflection beam. Overall, in terms of results, the SPR biosensor was much more effective in detection of exosomal PD-L1 in comparison to conventional methods like ELISA [93]. First and foremost, the study depicts that the SPR biosensor was very accurate and effective in detecting the PD-L1 in serum samples including that of a Stage III lung cancer patient; however, the method of ELISA was unable to detect PD-L1 in general. These serum samples were all 50 microliters. Additionally, through these biosensors, researchers proposed that the exosomal PD-L1 can be a biomarker (although this still needs to be confirmed through larger sample sets) for lung cancer diagnosis as the SPR biosensor detected much more exosomal PD-L1 in lung cancer patients compared to the normal controls [92]. The sensitivity and specificity of the SPR biosensor was much greater than the ELISA testing as the ELISA test could not detect any of the PD-L1 in serum samples in which there was a very low abundance of the exosomal PD-L1.

4.3. *Biosensors for Overall Cancer Immunotherapy*

Biosensors could play a pivotal role in augmenting cancer immunotherapy for treatment monitoring and evaluation of disease condition. For example, several study results revealed that the nano biosensor can be utilized cytokine secretion from individual cells for hours [94]. A specific biosensor is combined with a gold nanohole array sensor and a microfluidic-based system for accurate depiction and detection of cytokine secretion within a cell-to-cell basis. Additionally, some groups have worked on detection of the IL-2, IFN-γ, and TNF-α cytokines using gold nanorod-based biosensors [95]. Using these different types of sensors, researchers showed that immunoanalysis could be done on a smaller volume of analyte and a shorter timeframe than conventional ELISA methods [96]. Thus, these nanomaterial-based biosensors can be employed for more versatile applications than detection of a major immune checkpoint marker such as PD-L1.

4.4. *Biosensors for CRISPR*

Clustered regularly interspaced short palindromic repeats (CRISPR/Cas) based biosensors have also become more important for rapid detection of nucleic acids, exosomes, tumor DNAs, etc. CRISPR/Cas sensors use Cas effectors to cleave a specific DNA sequence. One of the main effectors, Cas9, has become well known due to its effective ability of editing genomes, identifying nucleic acids, and regulating transcription [97,98]. Specifically, Cas9 uses single guide RNA to help target a specific double-stranded DNA and cleave it for the purpose of selective genome editing [97,98]. In addition, the usage of nuclease-deactivated Cas9 can allow for binding and rebinding to the targeted double-stranded DNA, which is essential as fluorescent enzymes can be fused to the DNA to allow for detection of specific nucleic acid sequences due to the fluorescence and light emission [83,84]. This fusion of integrating Cas9 with a fluorescent enzyme was instrumental for researchers to detect DNA sequences of the American Zika virus. Other types of Cas effectors such as Cas12 and Cas13 have also been explored. Wherein, Cas12 effector is extremely important due to its ability to detect single-stranded DNA, which adds more to the detection of the different nucleic acids rather than focusing only double-stranded DNA with Cas9 effector. However, the main reason researchers have started to focus on CRISPR/Cas-based biosensors for cancer detection is due to the Cas13a effector. This effector works with the collateral cleavage activity of single-stranded RNAs. When there is a specific RNA of interest, the Cas13a effector cleaves RNA, the collateral cleavage of the reporter RNA (reRNA) which allows for the detection of the RNA with accurate quantification [85]. This effector is key for recognition of many miRNAs which can be either up-regulated or down-regulated. Hence, by being able to detect levels of miRNAs rapidly and effectively, this field of CRISPR/Cas-based biosensing is becoming a hot topic and displays promise for the early detection of cancer. However, one key limitation is that these biosensors involve sophisticated systems to run a tedious process to function fully [97,98]. Hence, reading these biosensors and their detection of the specific analyte requires well-trained professionals to understand the process and readout. Furthermore, the sensitivity of the biosensors is relatively low when compared to the other amplification methods [97,98]. Though many researchers are working on CRISPR/Cas-based biosensors, none of them could meet the needs of in vivo testing due to low sensitivity detection. Despite these issues, there is potential for the development of biosensor based on the Cas13a effector and the collateral cleavage activity detection [99].

4.5. Biosensors for ICB: Advantages and Limitations

Globally, the market size and demand for biosensors is rapidly increasing [98]. In 2026, the market size is expected to expand to about 37 billion dollars for all the different types of biosensors (electrochemical, FRET, piezoelectric, optical, etc.). However, improvements in biosensors are still a challenge because of the low concentration of biomarkers expressed over the cells. Nanomaterials-based detectors have the potential to amplify the signals with limit of detection at pico- or femtomolar concentration. In addition, the COVID-19 pandemic in 2020 has shown us the potential need of biosensors. Hence, in this review we have outlined current improvements and strategies in biosensors over the last decade for biomedical applications have been explored.

The detection of immune checkpoints via liquid biopsy presents an attractive option for early cancer detection. Currently, immune checkpoints are not widely used for diagnostics. Instead, they are utilized as prognostic markers for cancer and as predictors for the success of various immunotherapies. The biosensors proposed in recent years from the studies of Xu et al. [25] showcase the various advantages of using these biosensors for detecting small concentrations of PD-L1. For the quantification of different immune checkpoint markers like PD-L1, the studies showcased the limitations of ELISA and immunohistochemistry. For example, with ELISA, it is difficult to detect PD-L1 with very small concentrations. However, PD-L1 in multiple types of serum media is found in very small concentrations and thus, ELISA becomes extremely inaccurate in many situations. Immunohistochemistry is also an employed method that becomes very time-consuming and has very complex procedures. With these surface plasmon resonance-based biosensors and the multiplex biosensor, the studies show the simple-to-use nature of the biosensors, the very high specificity and sensitivity of the sensors, as well as their time and cost effectiveness. One way to ensure cost-effectiveness is a sensor that uses very cheap single-chain variable fragments from yeast cell lines in replacements of the very costly antibodies of the different immune checkpoint proteins that are used in the conventional methods.

Biosensors have been developed to detect the concentrations of the immune checkpoint markers in serum samples e.g., PD-L1 [42], but other potential IC markers can be identified through same approach. These biosensors demonstrated a few unique features with fine differences in the benefits and limitations of using one over the other. Tables 2 and 3 list the various advantages and disadvantages of these biosensors that are reported in the relevant literature [31,42,83,84,90]. Developments in surface plasmon resonance (SPR)-based biosensors development, especially for the detection of immune checkpoint markers in early detection of cancer, have progressed well due to several key advantages [31,92].

Table 3. Biomarkers and associated limits of detections (LOD) when using biosensors.

Cancer Type	Biomarker	Diagnostic (D) Prognostic (P)	LOD/mL	References {LOD References}
Tumor-Associated Antigens				
Breast	HER-2 ECD, CEA CA15-3	D/P	2 ng, 5 ng, 21.8U	[99–102] {[103–105]}
Ovarian	CA125, CA15-3	D	35 U, 12U	[21,106–109], {[106,108]}
Pancreatic	CEA	P	5 ng	[110] {[111]}
	CA19-9	D	37U	[112] {[113]}
	CA125	D/P	35U	[114,115] {[116]}
Gastric	CEA, HER-2 ECD	P	5 ng, 24.75 ng	[110,117] {[117,118]}
NSCLC	CEA	P	3–5ng	[119,120] {[120]}
Endometrial	CA 125	D/P	17.8U	[121,122] {[121]}
Colorectal	CEA	D/P	5ng	[123] {[124]}
Prostate	PSA	D	2.5–4 ng #	[125] {[126]}[127]
Immune Checkpoint Markers				
Primary central nervous system lymphoma (PCNSL)	PD-L1	D/P	0.43 ng	[78] {[79]}
NSCLC, DLBCL, CLL, NPC, HCC, ADR, PAC, HCC, Melanoma	PD-1	P	ND, 500 pg [1]	[73,74] {[73]}
Breast, Gastric	LAG-3	P	120 pg, 378.3 ng	[70,71] {[71,73]}
PAC, Osteosarcoma, Ovarian	TIM-3	P, D/P, ND	3 ng, 14.4 ng	[59,60,81,82] {[81,82]}

Note: ND: No optimal cutoff has been established; # Age dependent, non-small cell lung carcinoma (NSCLC), Diffuse large B-cell lymphoma (DLBCL), Chronic lymphocytic leukemia (CLL), Nasopharyngeal carcinoma (NPC), HCC, Advanced rectal cancer, Pancreatic adenocarcinoma (PAC), References in {} are corresponding to LOD values, [1] Melanoma.

LSPR biosensors show very high specificity and sensitivity (LOD: 1–5 pg/mL) for the immune checkpoint marker PD-L1 [31]. Furthermore, these biosensors are highly selective for the marker in comparison to the normal detection methods, e.g., immunohistochemistry, and can be integrated with a label-free detection system. Most of these sensors use fluorescent labeling or some other type of labeling approach to analyze the different analytes. However, the detection of analytes was required with different types of media which resulted in non-specific binding and high background signals. In addition, receptor labeling requires extensive time, effort, and optimization for the development of accurate biosensors for reproducible results. These limitations do limit the biosensors' usability. However, there is no additional labeling step required since this biosensor relies on physical properties of the analyte immobilized on a surface. Hence, these biosensors do not require nearly as much labor nor time to work effectively. Additionally, the biosensors allow for real time detection of the analyte as the analyte is localized into a compact surface; thus, there is no alteration of binding sites which could possibly occur with biosensors that require a labeling system.

SPR-based devices can cause configurational changes of the marker or analyte being detected by biosensors [32]. These biosensors use a mechanism of action where the analyte gets immobilized on the surface, often causing lower binding affinities due to the configurational changes of the immobilized receptor which, can cause the analyte binding ability and make it hard to detect the analyte accurately. SPR biosensors detect the marker or analyte by a change in wavelength with the aid of gold nanomaterials. Using these nanomaterials and immobilizing the analyte is a complex system to use for the commercialization of the product on a large scale.

5. Conclusions and Future Potential of Biosensors for ICB Detection

In this review, we have provided a detailed account on potential cancer markers and emerging IC markers, as well as their role in the tumor microenvironment (TME) in relation to cancer diagnosis. We presented the current state of IC markers' potential, in association with biosensors for the diagnosis and monitoring of various diseases and outlined the opportunities and applications of biosensors for improvements in healthcare. Thus, measuring and monitoring these IC markers in very minimal concentrations could assist in assessing early cancer. We have tabulated a list of biosensors applications in multiple studies and their limit of detections.

However, even though these biosensors have been proposed and built, it is important to note that only PD-L1 and CTLA-4 immune markers are currently approved for the treatment of ICB. Consequently, the future holds enormous potential for biosensor development to quantify other immune markers in cancerous microenvironments. Furthermore, surface plasmon resonance-based biosensors alongside nanoparticle usage are emerging as the most promising types of biosensors in quantification of these immune checkpoint markers. From other studies, the drawback of using these types of sensors is that they may be extremely complex to understand. Thus, a key objective, going forward, is to explore how to make these biosensors into user-friendly devices for cancer theranostics by using ICB.

In summary, in this review we clearly highlight the ongoing development of biosensors and their future potential for disease management, particularly for the early detection of cancer. In the future, these complex biosensors should be refined, as they should be able to measure with high sensitivity and simultaneous measurement of multiple biomarkers. In addition, it is imperative that the engineering and fabrication of biosensors should be easily accessible, low cost, and less complex to operate as a routine clinical testing tool for the detection of specific biomarkers. Furthermore, nanomaterial and nanoparticle-based biosensors have become prominent for early detection of cancer due to their high sensitivity as stated earlier in this review. However, additional focused investigations are required for the stability of these nanoparticles for prolonged periods with consistent and accurate readout.

Author Contributions: Conceptualization, A.N.; writing—original draft preparation, S.M., S.P., A.K.N., A.N.; writing—review and editing, S.M., S.P., A.K.N., A.N.; supervision, A.N.; project administration, A.N. All authors have read and agreed to the published version of the manuscript.

Funding: This research received no external funding.

Institutional Review Board Statement: Not applicable.

Informed Consent Statement: Not applicable.

Data Availability Statement: Not applicable.

Acknowledgments: The authors acknowledge the Molecular Imaging Program at Stanford and the Canary Center at Stanford for providing the support and Hema Brindha for assistance with drawing the figure.

Conflicts of Interest: The authors declare no conflict of interest.

Abbreviations

Ab_C	Capture antibody
Ab_s	Second antibody
ADAM	A disintegrin and metalloprotease
ALP	Alkaline phosphatase
BTLA	B- and T- lymphocyte attenuator
CEA	Carcinoembryonic antigen
CLL	Chronic lymphocytic leukemia
CRC	Colorectal cancer
CT scan	Computed tomography scan
CTC	Circulating tumor cells
ctDNA	Circulating tumor DNA
CTL	Cytotoxic T lymphocytes
CTLA-4	Cytotoxic T-lymphocyte–associated antigen 4
DLBCL	Diffuse large B-cell lymphoma
EC	Electrochemical biosensors
EGFR	Epidermal growth factor receptor
ELISA	Enzyme-linked immunoassay
FRET	Fluorescence resonance energy transfer
GSPE-MFD	Gold screen-printed electrode-based microfluidic devices
HCC	Hepatocellular carcinoma
HER2	Human epidermal growth factor receptor 2
Hq	Hydroquinone
HRP	Horse radish peroxide
IC	Immune checkpoints
ICB	Immue checkpoint blockade
IFN-γ	Interferon gamma
IgG	Immunoglobulin G
IHC	Immunohistochemistry
IL-2	Interleukin-2
LAG-3	Lymphocyte activation gene 3
LOD	Limit-of-detection
MAGE-A	Melanoma-associated antigen-A
MICB	Multiplexed immune checkpoint biosensor
MNP	Magnetic nanoparticles
MUC1	Mucin 1
NPC	Nasopharyngeal carcinoma
NTME	Non-tumor microenvironment
PD-1	Programmed death 1
PD-L1	Programmed death-1 ligand
PSA	Prostate specific antigen
PTC	Papillary thyroid cancer
ScFv	Single-chain variable fragment
sLAG-3	Serum lymphocyte activation gene-3
sPD-1	Serum programmed death-1
sPD-L1	Serum programmed death-1 ligand
SPR	Surface plasmon resonance
sTIM-3	Serum T cell immunoglobulin and mucin-domain containing-3
TAA	Tumor-associated antigen
TIM-3	T cell immunoglobulin and mucin-domain containing-3
TME	Tumor microenvironment
TNF-α	Tumor necrosis factor alpha
TSA	Tumor-specific antigen

References

1. Tothill, I.E. Biosensors for cancer markers diagnosis. *Semin. Cell Dev. Biol.* **2009**, *20*, 55–62. [CrossRef]
2. Cheng, N.; Du, D.; Wang, X.; Liu, D.; Xu, W.; Luo, Y. Recent Advances in Biosensors for Detecting Cancer-Derived Exosomes. *Trends Biotechnol.* **2019**, *37*, 1236–1254. [CrossRef]
3. Naresh, V.; Lee, N. A Review on Biosensors and Recent Development of Nanostructured Materials-Enabled Biosensors. *Sensors* **2021**, *21*, 1109. [CrossRef] [PubMed]
4. Mehrotra Biosensors and their applications—A review. *J. Oral. Biol. Craniofac. Res.* **2016**, *6*, 153–159. [CrossRef]
5. Soler, M.; Lechuga, L.M. Principles, technologies, and applications of plasmonic biosensors. *J. Appl. Physics* **2021**, *129*, 111102. [CrossRef]
6. Chen, Y.T.; Lee, Y.C.; Lai, Y.H.; Lim, J.C.; Huang, N.T.; Lin, C.T.; Huang, N.T. Review of Integrated Optical Biosensors for Point-Of-Care Applications. *Biosensors* **2020**, *10*, 209. [CrossRef]
7. Liu, D.; Wang, J.; Wu, L.; Huang, Y.; Zhang, Y.; Zhu, M.; Wang, Y.; Zhu, Z.; Yang, C. Trends in miniaturized biosensors for point-of-care testing. *TrAC Trends Anal. Chem.* **2020**, *122*, 115701. [CrossRef]
8. Dey, D.; Goswami, T. Optical biosensors: A revolution towards quantum nanoscale electronics device fabrication. *J. Biomed. Biotechnol.* **2011**, *2011*, 348218. [CrossRef] [PubMed]
9. Yi, J.; Xiao, W.; Li, G.; Wu, P.; He, Y.; Chen, C.; He, Y.; Ding, P.; Kai, T. The research of aptamer biosensor technologies for detection of microorganism. *Appl. MicroBiol. Biotechnol.* **2020**, *104*, 9877–9890. [CrossRef] [PubMed]
10. Chen, K.; Chou, W.; Liu, L.; Cui, Y.; Xue, P.; Jia, M. Electrochemical Sensors Fabricated by Electrospinning Technology: An Overview. *Sensors* **2019**, *19*, 3676. [CrossRef] [PubMed]
11. Park, S.; Boo, H.; Chung, T.D. Electrochemical non-enzymatic glucose sensors. *Anal. Chim. Acta* **2006**, *556*, 46–57. [CrossRef] [PubMed]
12. Kilgour, E.; Rothwell, D.G.; Brady, G.; Dive, C. Liquid Biopsy-Based Biomarkers of Treatment Response and Resistance. *Cancer Cell* **2020**, *37*, 485–495. [CrossRef] [PubMed]
13. Hollingsworth, R.E.; Jansen, K. Turning the corner on therapeutic cancer vaccines. *NPJ Vaccines* **2019**, *4*, 7. [CrossRef] [PubMed]
14. Briones, J.; Espulgar, W.; Koyama, S.; Takamatsu, H.; Tamiya, E.; Saito, M. The future of microfluidics in immune checkpoint blockade. *Cancer Gene Ther.* **2021**, *28*, 895–910. [CrossRef] [PubMed]
15. Theobald, M.; Biggs, J.; Dittmer, D.; Levine, A.J.; Sherman, L.A. Targeting p53 as a general tumor antigen. *Proc. Natl. Acad. Sci. USA* **1995**, *92*, 11993–11997. [CrossRef] [PubMed]
16. Casiano, C.A.; Mediavilla-Varela, M.; Tan, E.M. Tumor-associated antigen arrays for the serological diagnosis of cancer. *Mol. Cell Proteomics* **2006**, *5*, 1745–1759. [CrossRef]
17. Rangel, M.G.H.; Silva, M.L.S. Detection of the cancer-associated T antigen using an Arachis hypogea agglutinin biosensor. *Biosens Bioelectron.* **2019**, *141*, 111401. [CrossRef]
18. Kufe, D.W. MUC1-C oncoprotein as a target in breast cancer: Activation of signaling pathways and therapeutic approaches. *Oncogene* **2013**, *32*, 1073–1081. [CrossRef]
19. Iqbal, N.; Iqbal, N. Human Epidermal Growth Factor Receptor 2 (HER2) in Cancers: Overexpression and Therapeutic Implications. *Mol. Biol. Int.* **2014**, *2014*, 1–9. [CrossRef]
20. Gutierrez, C.; Schiff, R. HER2: Biology, detection, and clinical implications. *Arch. Pathol. Lab. Med.* **2011**, *135*, 55–62. [CrossRef] [PubMed]
21. Kallioniemi, O.P.; Kallioniemi, A.; Kurisu, W.; Thor, A.; Chen, L.C.; Smith, H.S.; Waldman, F.M.; Pinkel, D.; Gray, J.W. ERBB2 amplification in breast cancer analyzed by fluorescence in situ hybridization. *Proc. Natl. Acad. Sci. USA* **1992**, *89*, 5321–5325. [CrossRef]
22. Bakker, A.B.; Schreurs, M.W.; de Boer, A.J.; Kawakami, Y.; Rosenberg, S.A.; Adema, G.J.; Figdor, C.G. Melanocyte lineage-specific antigen gp100 is recognized by melanoma-derived tumor-infiltrating lymphocytes. *J. Exp. Med.* **1994**, *179*, 1005–1009. [CrossRef] [PubMed]
23. Cho, W.K.; Choi, D.H.; Park, H.C.; Park, W.; Yu, J.I.; Park, Y.S.; Lim, H.Y.; Kang, W.K.; Kim, H.C.; Cho, Y.B.; et al. Elevated CEA is associated with worse survival in recurrent rectal cancer. *Oncotarget* **2017**, *8*, 105936–105941. [CrossRef]
24. Tsikitis, V.L.; Malireddy, K.; Green, E.A.; Christensen, B.; Whelan, R.; Hyder, J.; Marcello, P.; Larach, S.; Lauter, D.; Sargent, D.J.; et al. Postoperative surveillance recommendations for early stage colon cancer based on results from the clinical outcomes of surgical therapy trial. *J. Clin. Oncol.* **2009**, *27*, 3671–3676. [CrossRef] [PubMed]
25. Xu, R.H.Z.; Liu, N.; Gao, F.; Luo, X. Electrochemical biosensors for the detection of carcinoembryonic antigenwith low fouling and high sensitivity based on copolymerized polydopamineand zwitterionic polymer. *Sens. Actuators B Chem.* **2020**, *319*, 128253. [CrossRef]
26. Gjerstorff, M.F.; Andersen, M.H.; Ditzel, H.J. Oncogenic cancer/testis antigens: Prime candidates for immunotherapy. *Oncotarget* **2015**, *6*, 15772–15787. [CrossRef]
27. Park, T.S.; Groh, E.M.; Patel, K.; Kerkar, S.P.; Lee, C.C.; Rosenberg, S.A. Expression of MAGE-A and NY-ESO-1 in Primary and Metastatic Cancers. *J. Immunother.* **2016**, *39*, 1–7. [CrossRef] [PubMed]
28. Salimi, A.; Kavosi, B.; Fathi, F.; Hallaj, R. Highly sensitive immunosensing of prostate-specific antigen based on ionic liquid-carbon nanotubes modified electrode: Application as cancer biomarker for prostate biopsies. *Biosens. Bioelectron.* **2013**, *42*, 439–446. [CrossRef] [PubMed]

29. Kavosi, B.; Salimi, A.; Hallaj, R.; Amani, K. A highly sensitive prostate-specific antigen immunosensor based on gold nanoparticles/PAMAM dendrimer loaded on MWCNTS/chitosan/ionic liquid nanocomposite. *Biosens. Bioelectron.* **2014**, *52*, 20–28. [CrossRef] [PubMed]
30. Damborska, D.; Bertok, T.; Dosekova, E.; Holazova, A.; Lorencova, L.; Kasak, P.; Tkac, J. Nanomaterial-based biosensors for detection of prostate specific antigen. *Mikrochim. Acta* **2017**, *184*, 3049–3067. [CrossRef] [PubMed]
31. Luo, B.; Wang, Y.; Lu, H.; Wu, S.; Lu, Y.; Shi, S.; Li, L.; Jiang, S.; Zhao, M. Label-free and specific detection of soluble programmed death ligand-1 using a localized surface plasmon resonance biosensor based on excessively tilted fiber gratings. *Biomed. Opt. Express* **2019**, *10*, 5136–5148. [CrossRef]
32. Helmerhorst, D.J.C.E.; Nussio, M.; Mamotte, C.D. Real-time and Label-free Bio-sensing of Molecular Interactions by Surface Plasmon Resonance: A Laboratory Medicine Perspective. *Clin. Biochem. Rev.* **2012**, *33*, 161–173.
33. Luo, B.; Wang, Y.; Lu, H.; Wu, S.; Lu, Y.; Shi, S.; Li, L.; Jiang, S.; Zhao, M. Localized surface plasmon resonance based nanobiosensor for biomarker detection of invasive cancer cells. *J. Biomed. Opt.* **2014**, *19*, 51202.
34. Chen, J.S.; Chen, P.F.; Lin, H.T.; Huang, N.T. A Localized surface plasmon resonance (LSPR) sensor integrated automated microfluidic system for multiplex inflammatory biomarker detection. *Analyst* **2020**, *145*, 7654–7661. [CrossRef]
35. Khanmohammadi, A.; Aghaie, A.; Vahedi, E.; Qazvini, A.; Ghanei, M.; Afkhami, A.; Hajian, A.; Bagheri, H. Electrochemical biosensors for the detection of lung cancer biomarkers: A review. *Talanta* **2020**, *206*, 120251. [CrossRef]
36. Jalil, O.; Pandey, C.M.; Kumar, D. Electrochemical biosensor for the epithelial cancer biomarker EpCAM based on reduced graphene oxide modified with nanostructured titanium dioxide. *Mikrochim. Acta* **2020**, *187*, 275. [CrossRef] [PubMed]
37. El Aamri, M.; Yammouri, G.; Mohammadi, H.; Amine, A.; Korri-Youssoufi, H. Electrochemical Biosensors for Detection of MicroRNA as a Cancer Biomarker: Pros and Cons. *Biosensors* **2020**, *10*, 186. [CrossRef]
38. Gajdosova, V.; Lorencova, L.; Kasak, P.; Tkac, J. Electrochemical Nanobiosensors for Detection of Breast Cancer Biomarkers. *Sensors* **2020**, *20*, 4022. [CrossRef] [PubMed]
39. Xiao, L.; Zhu, A.; Xu, Q.; Chen, Y.; Xu, J.; Weng, J. Colorimetric Biosensor for Detection of Cancer Biomarker by Au Nanoparticle-Decorated Bi2Se3 Nanosheets. *ACS Appl. Mater. Interfaces* **2017**, *9*, 6931–6940. [CrossRef]
40. Cai, J.; Ding, L.; Gong, P.; Huang, J. A colorimetric detection of microRNA-148a in gastric cancer by gold nanoparticle-RNA conjugates. *Nanotechnology* **2020**, *31*, 95501. [CrossRef] [PubMed]
41. Mollasalehi, H.; Shajari, E. A colorimetric nano-biosensor for simultaneous detection of prevalent cancers using unamplified cell-free ribonucleic acid biomarkers. *Bioorg. Chem.* **2021**, *107*, 104605. [CrossRef]
42. Wuethrich, A.; Rajkumar, A.R.; Shanmugasundaram, K.B.; Reza, K.K.; Dey, S.; Howard, C.B.; Sina, A.A.; Trau, M. Single droplet detection of immune checkpoints on a multiplexed electrohydrodynamic biosensor. *Analyst* **2019**, *144*, 6914–6921. [CrossRef]
43. Cho, S.; Yang, H.C.; Rhee, W.J. Simultaneous multiplexed detection of exosomal microRNAs and surface proteins for prostate cancer diagnosis. *Biosens. Bioelectron.* **2019**, *146*, 111749. [CrossRef]
44. Wei, X.; Bian, F.; Cai, X.; Wang, Y.; Cai, L.; Yang, J.; Zhu, Y.; Zhao, Y. Multiplexed Detection Strategy for Bladder Cancer MicroRNAs Based on Photonic Crystal Barcodes. *Anal. Chem.* **2020**, *92*, 6121–6127. [CrossRef]
45. Traynor, S.M. Recent Advances in Electrochemical Detection of Prostate Specific Antigen (PSA) in Clinically-Relevant Samples. *J. Electrochem. Soc.* **2020**, *167*, 037551. [CrossRef]
46. Li, H.; Li, S.; Xia, F. *Biosensors Based on Sandwich Assays*; Springer: Berlin/Heidelberg, Germany, 2018.
47. Neupane, D.; Stine, K.J. Electrochemical Sandwich Assays for Biomarkers Incorporating Aptamers, Antibodies and Nanomaterials for Detection of Specific Protein Biomarkers. *Appl. Sci.* **2021**, *11*, 7087. [CrossRef]
48. Chen, S.; Wang, Z.; Cui, X.; Jiang, L.; Zhi, Y.; Ding, X.; Nie, Z.; Zhou, P.; Cui, D. Microfluidic Device Directly Fabricated on Screen-Printed Electrodes for Ultrasensitive Electrochemical Sensing of PSA. *Nanoscale Res. Lett.* **2019**, *14*, 71. [CrossRef] [PubMed]
49. Zani, A.; Laschi, S.; Mascini, M.; Marrazza, G. A New Electrochemical Multiplexed Assay for PSA Cancer Marker Detection. *Electroanalysis* **2011**, *23*, 91. [CrossRef]
50. Niu, X.; Cheng, N.; Ruan, X.; Du, D.; Lin, Y. Nanozyme-Based Immunosensors and Immunoassays: Recent Developments and Future Trends. *J. Electrochem. Soc.* **2020**, *167*, 37508. [CrossRef]
51. Robert, C. A decade of immune-checkpoint inhibitors in cancer therapy. *Nat. Commun.* **2020**, *11*, 3801. [CrossRef] [PubMed]
52. Robert, C.; Ribas, A.; Hamid, O.; Daud, A.; Wolchok, J.D.; Joshua, A.M.; Hwu, W.J.; Weber, J.S.; Gangadhar, T.C.; Joseph, R.W.; et al. Durable Complete Response After Discontinuation of Pembrolizumab in Patients with Metastatic Melanoma. *J. Clin. Oncol.* **2018**, *36*, 1668–1674. [CrossRef]
53. Vaddepally, R.K.; Kharel, P.; Pandey, R.; Garje, R.; Chandra, A.B. Review of Indications of FDA-Approved Immune Checkpoint Inhibitors per NCCN Guidelines with the Level of Evidence. *Cancers* **2020**, *12*, 738. [CrossRef]
54. Qin, S.; Xu, L.; Yi, M.; Yu, S.; Wu, K.; Luo, S. Novel immune checkpoint targets: Moving beyond PD-1 and CTLA-4. *Mol. Cancer* **2019**, *18*, 155. [CrossRef]
55. Shibru, B.; Fey, K.; Fricke, S.; Blaudszun, A.R.; Furst, F.; Weise, M.; Seiffert, S.; Weyh, M.K.; Kohl, U.; Sack, U.; et al. Detection of Immune Checkpoint Receptors—A Current Challenge in Clinical Flow Cytometry. *Front. Immunol.* **2021**, *12*, 694055. [CrossRef]
56. Wolf, Y.; Anderson, A.C.; Kuchroo, V.K. TIM3 comes of age as an inhibitory receptor. *Nat. Rev. Immunol.* **2020**, *20*, 173–185. [CrossRef]
57. Acharya, N.; Sabatos-Peyton, C.; Anderson, A.C. Tim-3 finds its place in the cancer immunotherapy landscape. *J. Immunother. Cancer* **2020**, *8*, e000911. [CrossRef] [PubMed]

58. Banerjee, H.; Kane, L.P. Immune regulation by Tim-3. *F1000 Res.* **2018**, *7*, 316. [CrossRef] [PubMed]
59. Wu, J.; Liu, C.; Qian, S.; Hou, H. The expression of Tim-3 in peripheral blood of ovarian cancer. *DNA Cell Biol.* **2013**, *32*, 648–653. [CrossRef] [PubMed]
60. Liu, H.; Zhi, L.; Duan, N.; Su, P. Abnormal expression of Tim-3 antigen on peripheral blood T cells is associated with progressive disease in osteosarcoma patients. *FEBS Open Biol.* **2016**, *6*, 807–815. [CrossRef] [PubMed]
61. Chen, F.; Zhang, X.H.; Hu, X.D.; Liu, P.D.; Zhang, H.Q. The effects of combined selenium nanoparticles and radiation therapy on breast cancer cells in vitro. *Artif. Cells Nanomed. Biotechnol.* **2018**, *46*, 937–948. [CrossRef] [PubMed]
62. Yu, X.; Zheng, Y.; Mao, R.; Su, Z.; Zhang, J. BTLA/HVEM Signaling: Milestones in Research and Role in Chronic Hepatitis B Virus Infection. *Front. Immunol.* **2019**, *10*, 617. [CrossRef]
63. Deng, Z.; Zheng, Y.; Cai, P.; Zheng, Z. The Role of B and T Lymphocyte Attenuator in Respiratory System Diseases. *Front. Immunol.* **2021**, *12*, 635623. [CrossRef]
64. Lan, X.; Li, S.; Gao, H.; Nanding, A.; Quan, L.; Yang, C.; Ding, S.; Xue, Y. Increased BTLA and HVEM in gastric cancer are associated with progression and poor prognosis. *Onco Targets Ther.* **2017**, *10*, 919–926. [CrossRef]
65. König, B.; Grätzel, M. An immunosensor for the detection of human B-lymphocytes. *Bioorg. Med. Chem. Lett.* **1994**, *4*, 2429–2434. [CrossRef]
66. Carinelli, S.; Ballesteros, C.X.; Marti, M.; Alegret, S.; Pividori, M.I. Electrochemical magneto-actuated biosensor for CD4 count in AIDS diagnosis and monitoring. *Biosens. Bioelectron.* **2015**, *74*, 974–980. [CrossRef]
67. Kim, J.; Park, G.; Lee, S.; Hwang, S.W.; Min, N.; Lee, K.M. Single wall carbon nanotube electrode system capable of quantitative detection of CD4(+) T cells. *Biosens. Bioelectron.* **2017**, *90*, 238–244. [CrossRef] [PubMed]
68. Wang, H.; Cai, H.H.; Zhang, L.; Cai, J.; Yang, P.H.; Chen, Z.W. A novel gold nanoparticle-doped polyaniline nanofibers-based cytosensor confers simple and efficient evaluation of T-cell activation. *Biosens. Bioelectron.* **2013**, *50*, 167–173. [CrossRef]
69. Andrews, L.P.; Marciscano, A.E.; Drake, C.G.; Vignali, D.A. LAG3 (CD223) as a cancer immunotherapy target. *Immunol. Rev.* **2017**, *276*, 80–96. [CrossRef] [PubMed]
70. He, Y.; Wang, Y.; Zhao, S.; Zhao, C.; Zhou, C.; Hirsch, F.R. sLAG-3 in non-small-cell lung cancer patients' serum. *Oncol. Targets Ther.* **2018**, *11*, 4781–4784. [CrossRef] [PubMed]
71. Li, N.; Jilisihan, B.; Wang, W.; Tang, Y.; Keyoumu, S. Soluble LAG3 acts as a potential prognostic marker of gastric cancer and its positive correlation with CD8+T cell frequency and secretion of IL-12 and INF-gamma in peripheral blood. *Cancer Biomark* **2018**, *23*, 341–351. [CrossRef]
72. Triebel, F.; Hacene, K.; Pichon, M.F. A soluble lymphocyte activation gene-3 (sLAG-3) protein as a prognostic factor in human breast cancer expressing estrogen or progesterone receptors. *Cancer Lett.* **2006**, *235*, 147–153. [CrossRef]
73. Ugurel, S.; Schadendorf, D.; Horny, K.; Sucker, A.; Schramm, S.; Utikal, J.; Pfohler, C.; Herbst, R.; Schilling, B.; Blank, C.; et al. Elevated baseline serum PD-1 or PD-L1 predicts poor outcome of PD-1 inhibition therapy in metastatic melanoma. *Ann. Oncol.* **2020**, *31*, 144–152. [CrossRef]
74. Khan, M.; Zhao, Z.; Arooj, S.; Fu, Y.; Liao, G. Soluble PD-1: Predictive, Prognostic, and Therapeutic Value for Cancer Immunotherapy. *Front. Immunol.* **2020**, *11*, 587460. [CrossRef]
75. Li, Y.; Cui, X.; Yang, Y.J.; Chen, Q.Q.; Zhong, L.; Zhang, T.; Cai, R.L.; Miao, J.Y.; Yu, S.C.; Zhang, F. Serum sPD-1 and sPD-L1 as Biomarkers for Evaluating the Efficacy of Neoadjuvant Chemotherapy in Triple-Negative Breast Cancer Patients. *Clin. Breast Cancer* **2019**, *19*, 326–332.e1. [CrossRef]
76. Romero, Y.; Wise, R.; Zolkiewska, A. Proteolytic processing of PD-L1 by ADAM proteases in breast cancer cells. *Cancer Immunol. Immunother.* **2020**, *69*, 43–55. [CrossRef]
77. Aghajani, M.J.; Roberts, T.L.; Yang, T.; McCafferty, C.E.; Caixeiro, N.J.; De Souza, P.; Niles, N. Elevated levels of soluble PD-L1 are associated with reduced recurrence in papillary thyroid cancer. *Endocr. Connect.* **2019**, *8*, 1040–1051. [CrossRef] [PubMed]
78. Cho, I.; Lee, H.; Yoon, S.E.; Ryu, K.J.; Ko, Y.H.; Kim, W.S.; Kim, S.J. Serum levels of soluble programmed death-ligand 1 (sPD-L1) in patients with primary central nervous system diffuse large B-cell lymphoma. *BMC Cancer* **2020**, *20*, 120. [CrossRef] [PubMed]
79. Moller-Hackbarth, K.; Dewitz, C.; Schweigert, O.; Trad, A.; Garbers, C.; Rose-John, S.; Scheller, J. A disintegrin and metalloprotease (ADAM) 10 and ADAM17 are major sheddases of T cell immunoglobulin and mucin domain 3 (Tim-3). *J. Biol. Chem.* **2013**, *288*, 34529–34544. [CrossRef]
80. Ge, W.; Li, J.; Fan, W.; Xu, D.; Sun, S. Tim-3 as a diagnostic and prognostic biomarker of osteosarcoma. *Tumour Biol.* **2017**, *39*, 1010428317715643. [CrossRef] [PubMed]
81. Li, F.; Li, N.; Sang, J.; Fan, X.; Deng, H.; Zhang, X.; Han, Q.; Lv, Y.; Liu, Z. Highly elevated soluble Tim-3 levels correlate with increased hepatocellular carcinoma risk and poor survival of hepatocellular carcinoma patients in chronic hepatitis B virus infection. *Cancer Manag Res.* **2018**, *10*, 941–951. [CrossRef]
82. Kangas, M.J.; Burks, R.M.; Atwater, J.; Lukowicz, R.M.; Williams, P.; Holmes, A.E. Colorimetric Sensor Arrays for the Detection and Identification of Chemical Weapons and Explosives. *Crit. Rev. Anal. Chem.* **2017**, *47*, 138–153. [CrossRef]
83. Menon, S.; Mathew, M.R.; Sam, S.; Keerthi, K.; Kumar, K.G. Recent advances and challenges in electrochemical biosensors for emerging and re-emerging infectious diseases. *J. Electroana.l Chem.* **2020**, *878*, 114596. [CrossRef] [PubMed]
84. Alsaab, H.O.; Sau, S.; Alzhrani, R.; Tatiparti, K.; Bhise, K.; Kashaw, S.K.; Iyer, A.K. PD-1 and PD-L1 Checkpoint Signaling Inhibition for Cancer Immunotherapy: Mechanism, Combinations, and Clinical Outcome. *Front. Pharmacol.* **2017**, *8*, 561. [CrossRef]

85. Jiang, Y.; Chen, M.; Nie, H.; Yuan, Y. PD-1 and PD-L1 in cancer immunotherapy: Clinical implications and future considerations. *Hum. Vaccin. Immunother.* **2019**, *15*, 1111–1122. [CrossRef]
86. Kruger, S.; Legenstein, M.L.; Rosgen, V.; Haas, M.; Modest, D.P.; Westphalen, C.B.; Ormanns, S.; Kirchner, T.; Heinemann, V.; Holdenrieder, S.; et al. Serum levels of soluble programmed death protein 1 (sPD-1) and soluble programmed death ligand 1 (sPD-L1) in advanced pancreatic cancer. *Oncoimmunology* **2017**, *6*, e1310358. [CrossRef]
87. Cunningham, R.A.; Holland, M.; McWilliams, E.; Hodi, F.S.; Severgnini, M. Detection of clinically relevant immune checkpoint markers by multicolor flow cytometry. *J. Biol. Methods* **2019**, *6*, e114. [CrossRef]
88. Liu, R.; Ye, X.; Cui, T. Recent Progress of Biomarker Detection Sensors. *Research* **2020**, *2020*, 7949037. [CrossRef] [PubMed]
89. Azmi, A.S.; Bao, B.; Sarkar, F.H. Exosomes in cancer development, metastasis, and drug resistance: A comprehensive review. *Cancer Metastasis Rev.* **2013**, *32*, 623–642. [CrossRef]
90. Liu, C.; Zeng, X.; An, Z.; Yang, Y.; Eisenbaum, M.; Gu, X.; Jornet, J.M.; Dy, G.K.; Reid, M.E.; Gan, Q.; et al. Sensitive Detection of Exosomal Proteins via a Compact Surface Plasmon Resonance Biosensor for Cancer Diagnosis. *ACS Sens.* **2018**, *3*, 1471–1479. [CrossRef] [PubMed]
91. Cordonnier, M.; Nardin, C.; Chanteloup, G.; Derangere, V.; Algros, M.P.; Arnould, L.; Garrido, C.; Aubin, F.; Gobbo, J. Tracking the evolution of circulating exosomal-PD-L1 to monitor melanoma patients. *J. Extracell Vesicles* **2020**, *9*, 1710899. [CrossRef]
92. Li, X.; Soler, M.; Szydzik, C.; Khoshmanesh, K.; Schmidt, J.; Coukos, G.; Mitchell, A.; Altug, H. Label-Free Optofluidic Nanobiosensor Enables Real-Time Analysis of Single-Cell Cytokine Secretion. *Small* **2018**, *14*, e1800698. [CrossRef]
93. Dutta, N.; Lillehoj, P.B.; Estrela, P.; Dutta, G. Electrochemical Biosensors for Cytokine Profiling: Recent Advancements and Possibilities in the Near Future. *Biosensors* **2021**, *11*, 94. [CrossRef]
94. Oh, B.R.; Huang, N.T.; Chen, W.; Seo, J.H.; Chen, P.; Cornell, T.T.; Shanley, T.P.; Fu, J.; Kurabayashi, K. Integrated nanoplasmonic sensing for cellular functional immunoanalysis using human blood. *ACS Nano* **2014**, *8*, 2667–2676. [CrossRef]
95. Li, Y.; Li, S.; Wang, J.; Liu, G. CRISPR/Cas Systems towards Next-Generation Biosensing. *Trends Biotechnol.* **2019**, *37*, 730–743. [CrossRef]
96. Mukama, O.; Wu, J.; Li, Z.; Liang, Q.; Yi, Z.; Lu, X.; Liu, Y.; Liu, Y.; Hussain, M.; Makafe, G.G.; et al. An ultrasensitive and specific point-of-care CRISPR/Cas12 based lateral flow biosensor for the rapid detection of nucleic acids. *Biosens. Bioelectron.* **2020**, *159*, 112143. [CrossRef]
97. Bruch, R.; Baaske, J.; Chatelle, C.; Meirich, M.; Madlener, S.; Weber, W.; Dincer, C.; Urban, G.A. CRISPR/Cas13a-Powered Electrochemical Microfluidic Biosensor for Nucleic Acid Amplification-Free miRNA Diagnostics. *Adv. Mater.* **2019**, *31*, e1905311. [CrossRef] [PubMed]
98. Sadana, A. Market Size and Economics for Biosensors. In *Fractal Binding and Dissociation Kinetics for Different Biosensor Applications*; Elsevier: Amsterdam, The Netherlands, 2005; pp. 265–299.
99. Tchou, J.; Lam, L.; Li, Y.R.; Edwards, C.; Ky, B.; Zhang, H. Monitoring serum HER2 levels in breast cancer patients. *Springerplus* **2015**, *4*, 237. [CrossRef] [PubMed]
100. Lee, M.H.; Jung, S.Y.; Kang, S.H.; Song, E.J.; Park, I.H.; Kong, S.Y.; Kwon, Y.M.; Lee, K.S.; Kang, H.S.; Lee, E.S. The Significance of Serum HER2 Levels at Diagnosis on Intrinsic Subtype-Specific Outcome of Operable Breast Cancer Patients. *PLoS ONE* **2016**, *11*, e0163370. [CrossRef]
101. Leyland-Jones, B.; Smith, B.R. Serum HER2 testing in patients with HER2-positive breast cancer: The death knell tolls. *Lancet Oncol.* **2011**, *12*, 286–295. [CrossRef]
102. Lee, J.S.; Park, S.; Park, J.M.; Cho, J.H.; Kim, S.I.; Park, B.W. Elevated levels of serum tumor markers CA 15-3 and CEA are prognostic factors for diagnosis of metastatic breast cancers. *Breast Cancer Res. Treat.* **2013**, *141*, 477–484. [CrossRef]
103. Fakhari, A.; Gharepapagh, E.; Dabiri, S.; Gilani, N. Correlation of cancer antigen 15-3 (CA15-3) serum level and bony metastases in breast cancer patients. *Med. J. Islam Repub. Iran* **2019**, *33*, 142. [CrossRef] [PubMed]
104. Shao, Y.; Sun, X.; He, Y.; Liu, C.; Liu, H. Elevated Levels of Serum Tumor Markers CEA and CA15-3 Are Prognostic Parameters for Different Molecular Subtypes of Breast Cancer. *PLoS ONE* **2015**, *10*, e0133830. [CrossRef] [PubMed]
105. Loo, L.; Capobianco, J.A.; Wu, W.; Gao, X.; Shih, W.Y.; Shih, W.H.; Pourrezaei, K.; Robinson, M.K.; Adams, G.P. Highly sensitive detection of HER2 extracellular domain in the serum of breast cancer patients by piezoelectric microcantilevers. *Anal. Chem.* **2011**, *83*, 3392–3397. [CrossRef]
106. Scambia, G.; Panici, P.B.; Baiocchi, G.; Perrone, L.; Greggi, S.; di Roberto, P.; Mancuso, S. CA 15-3 serum levels in ovarian cancer. *Oncology* **1988**, *45*, 263–267. [CrossRef]
107. Scambia, G.; Panici, P.B.; Baiocchi, G.; Perrone, L.; Greggi, S.; Mancuso, S. CA 15-3 as a tumor marker in gynecological malignancies. *Gynecol. Oncol.* **1988**, *30*, 265–273. [CrossRef]
108. Charkhchi, P.; Cybulski, C.; Gronwald, J.; Wong, F.O.; Narod, S.A.; Akbari, M.R. CA125 and Ovarian Cancer: A Comprehensive Review. *Cancers* **2020**, *12*, 3730. [CrossRef]
109. Dochez, V.; Caillon, H.; Vaucel, E.; Dimet, J.; Winer, N.; Ducarme, G. Biomarkers and algorithms for diagnosis of ovarian cancer: CA125, HE4, RMI and ROMA, a review. *J. Ovarian Res.* **2019**, *12*, 28. [CrossRef] [PubMed]
110. Nan, J.; Li, J.; Li, X.; Guo, G.; Wen, X.; Tian, Y. Preoperative Serum Carcinoembryonic Antigen as a Marker for Predicting the Outcome of Three Cancers. *Biomark. Cancer* **2017**, *9*, 1–7. [CrossRef]

111. Meng, Q.; Shi, S.; Liang, C.; Liang, D.; Xu, W.; Ji, S.; Zhang, B.; Ni, Q.; Xu, J.; Yu, X. Diagnostic and prognostic value of carcinoembryonic antigen in pancreatic cancer: A systematic review and meta-analysis. *Oncol. Targets Ther.* **2017**, *10*, 4591–4598. [CrossRef]
112. Ballehaninna, U.K.; Chamberlain, R.S. Serum CA 19-9 as a Biomarker for Pancreatic Cancer-A Comprehensive Review. *Indian J. Surg. Oncol.* **2011**, *2*, 88–100. [CrossRef]
113. Poruk, K.E.; Gay, D.Z.; Brown, K.; Mulvihill, J.D.; Boucher, K.M.; Scaife, C.L.; Firpo, M.A.; Mulvihill, S.J. The clinical utility of CA 19-9 in pancreatic adenocarcinoma: Diagnostic and prognostic updates. *Curr. Mol. Med.* **2013**, *13*, 340–351.
114. Haglund, C. Tumour marker antigen CA125 in pancreatic cancer: A comparison with CA19-9 and CEA. *Br. J. Cancer* **1986**, *54*, 897–901. [CrossRef]
115. Liu, L.; Xu, H.X.; Wang, W.Q.; Wu, C.T.; Xiang, J.F.; Liu, C.; Long, J.; Xu, J.; Fu, L.; Ni, Q.X. Serum CA125 is a novel predictive marker for pancreatic cancer metastasis and correlates with the metastasis-associated burden. *Oncotarget* **2016**, *7*, 5943–5956. [CrossRef]
116. Penney, D.G.; Weeks, T.A. Age dependence of cardiac growth in the normal and carbon monoxide-exposed rat. *Dev. Biol.* **1979**, *71*, 153–162. [CrossRef]
117. Shi, H.Z.; Wang, Y.N.; Huang, X.H.; Zhang, K.C.; Xi, H.Q.; Cui, J.X.; Liu, G.X.; Liang, W.T.; Wei, B.; Chen, L. Serum HER2 as a predictive biomarker for tissue HER2 status and prognosis in patients with gastric cancer. *World J. Gastroenterol.* **2017**, *23*, 1836–1842. [CrossRef]
118. Cainap, C.; Nagy, V.; Gherman, A.; Cetean, S.; Laszlo, I.; Constantin, A.M.; Cainap, S. Classic tumor markers in gastric cancer. Current standards and limitations. *Clujul Med.* **2015**, *88*, 111–115.
119. Arrieta, O.; Villarreal-Garza, C.; Martinez-Barrera, L.; Morales, M.; Dorantes-Gallareta, Y.; Pena-Curiel, O.; Contreras-Reyes, S.; Macedo-Perez, E.O.; Alatorre-Alexander, J. Usefulness of serum carcinoembryonic antigen (CEA) in evaluating response to chemotherapy in patients with advanced non-small-cell lung cancer: A prospective cohort study. *BMC Cancer* **2013**, *13*, 254. [CrossRef]
120. Grunnet, M.; Sorensen, J.B. Carcinoembryonic antigen (CEA) as tumor marker in lung cancer. *Lung Cancer* **2012**, *76*, 138–143. [CrossRef]
121. Nithin, K.U.; Sridhar, M.G.; Srilatha, K.; Habebullah, S. CA 125 is a better marker to differentiate endometrial cancer and abnormal uterine bleeding. *Afr. Health Sci.* **2018**, *18*, 972–978.
122. Duk, J.M.; Aalders, J.G.; Fleuren, G.J.; de Bruijn, H.W.A. CA 125: A useful marker in endometrial carcinoma. *Am. J. Obst. Gynecol.* **1986**, *155*, 1097–1102. [CrossRef]
123. Goldstein, M.J.; Mitchell, E.P. Carcinoembryonic antigen in the staging and follow-up of patients with colorectal cancer. *Cancer Invest.* **2005**, *23*, 338–351. [CrossRef] [PubMed]
124. Su, B.B.; Shi, H.; Wan, J. Role of serum carcinoembryonic antigen in the detection of colorectal cancer before and after surgical resection. *World J. Gastroenterol.* **2012**, *18*, 2121–2126. [CrossRef] [PubMed]
125. Catalona, W.J.; Smith, D.S.; Ornstein, D.K. Prostate cancer detection in men with serum PSA concentrations of 2.6 to 4.0 ng/mL and benign prostate examination. Enhancement of specificity with free PSA measurements. *JAMA* **1997**, *277*, 1452–1455. [CrossRef]
126. Catalona, W.J.; Loeb, S. Prostate cancer screening and determining the appropriate prostate-specific antigen cutoff values. *J. Natl. Compr. Cancer Netw.* **2010**, *8*, 265–270. [CrossRef] [PubMed]
127. Benvenuto, M.; Focaccetti, C.; Izzi, V.; Masuelli, L.; Modesti, A.; Bei, R. Tumor antigens heterogeneity and immune response-targeting neoantigens in breast cancer. *Semin Cancer Biol.* **2021**, *72*, 65–75. [CrossRef] [PubMed]

 biosensors

Article

Highly Sensitive Detection and Differentiation of Endotoxins Derived from Bacterial Pathogens by Surface-Enhanced Raman Scattering

Xiaomeng Wu [1], Yiping Zhao [2,*] and Susu M. Zughaier [3,4,*]

1 College of Food Science and Nutritional Engineering, National Engineering Research Center for Fruit & Vegetable Processing, Key Laboratory of Fruit & Vegetable Processing, Ministry of Agriculture, Beijing Key Laboratory for Food Non-Thermal Processing, China Agricultural University, Beijing 100083, China; wuxmeng@cau.edu.cn
2 Department of Physics and Astronomy, University of Georgia, Athens, GA 30602, USA
3 Department of Basic Medical Sciences, College of Medicine, QU Health, Qatar University, Doha P.O. Box 2731, Qatar
4 Biomedical and Pharmaceutical Research Unit, QU Health, Qatar University, Doha P.O. Box 2731, Qatar
* Correspondence: zhaoy@uga.edu (Y.Z.); szughaier@qu.edu.qa (S.M.Z.)

Abstract: Bacterial endotoxins, as major components of Gram-negative bacterial outer membrane leaflets and a well-characterized TLR4-MD-2 ligand, are lipopolysaccharides (LPSs) that are constantly shed from bacteria during growth and infection. For the first time, we report that unique surface-enhanced Raman scattering (SERS) spectra of enteric LPSs from *E. coli*, *S. typhimurium*, *S. minnesota*, *V. cholerae*, *Rhizobium* species *R.* CE3, and *R.* NGR, as well as *Neisseria meningitidis* endotoxin structures, LPSs, lipid A, and KDO2-lipid A can be obtained. The characteristic peaks of the SERS spectra reveal that most of the tested LPS structures are from lipids and saccharides, i.e., the major components of LPSs, and these spectra can be successfully used to differentiate between endotoxins with principal components analysis. In addition, all the LPS samples here are measured at a concentration of 10 nmole/mL, which corresponds to their relevant pathophysiological concentrations in clinical infections. This study demonstrates that LPSs can be used as biomarkers for the highly sensitive detection of bacteria using SERS-based methods.

Keywords: SERS; LPS; bacteria; endotoxin; lipid A; silver nanorods

Citation: Wu, X.; Zhao, Y.; Zughaier, S.M. Highly Sensitive Detection and Differentiation of Endotoxins Derived from Bacterial Pathogens by Surface-Enhanced Raman Scattering *Biosensors* **2021**, *11*, 234. https://doi.org/10.3390/bios11070234

Received: 12 May 2021
Accepted: 2 June 2021
Published: 11 July 2021

Publisher's Note: MDPI stays neutral with regard to jurisdictional claims in published maps and institutional affiliations.

1. Introduction

An endotoxin or lipopolysaccharide is a glycolipid and is a major component of the outer membrane in Gram-negative bacteria. Endotoxins are shed from live bacteria as membrane blebs and vesicles or released from dead bacteria into tissue at the site of infection. LPSs are well-characterized pathogen-associated molecular pattern (PAMP) ligands that bind to the human TLR4-MD-2 receptor and elicit strong proinflammatory responses in immune cells [1,2]. For example, *Neisseria meningitidis* is a strictly human pathogen that causes meningitis and is the leading cause of fulminant sepsis and death [3]. The meningococcal LPS molecule produced by *Neisseria meningitidis* is composed of a lipid A component containing di-glucosamine that is linked to two KDO sugars and a heptose containing sugar or saccharide chain, as shown in Figure A1. The meningococcal LPS is a very potent inducer of proinflammatory mediator release, known as the cytokine storm, which contributes to massive fulminant meningococcal sepsis and rapid death [3]. As such, the rapid and sensitive detection of trace amounts of meningococcal LPSs in biological fluids is a highly desirable approach to help diagnose meningococcal infection, thus facilitating therapy and saving lives.

Similarly, other Gram-negative bacteria contain endotoxins or LPSs as a major component of their outer membrane. Enteric bacteria LPS structures also contain a lipid A moiety linked to repeating units of polysaccharide chains, known as the O-antigen. Similar to meningococcal LPSs, enteric KDO_2-lipid A is responsible for the endotoxic activity of LPSs; however, LPS and lipid A structures from enteric pathogens are more diverse and vary greatly in their ability to elicit TLR4-MD-2 mediated inflammatory responses [4]. Enteric lipid A structures, such as *E. coli*, *Salmonella*, and *Klebsiella*, all have fatty acyl chain lengths ranging from 10 to 16 carbon atoms, as shown in Figure A2, while other Gram-negative bacteria like *Bacteroides* and *Rhizobium* contain branched fatty acyl chains with extended lengths of up to 28 carbon atoms [5]. The length of the fatty acyl chain impacts lipid A endotoxic activity, where extended length fatty acyl chains dramatically reduce lipid A potency and act as a TLR4-MD-2 antagonist rather than as an agonist. These pathogenic Gram-negative bacteria can shed active endotoxins and elicit pro-inflammatory responses. Consequently, the rapid and sensitive detection of enteric endotoxin structures in biological fluids and consumed food would greatly restrict the transmission of infections.

The current methods to detect endotoxins in biological fluids and environmental samples are often time-consuming and costly, and some methods are not sensitive enough. A limulus amebocyte lysate (LAL) assay is the most common method to detect an endotoxin [6]. The LAL assay depends on the enzyme purified from a horseshoe crab that forms a clot upon detecting an endotoxin [6]. In spite of the significant enhancement of enzyme specificity and LAL assay performance, the method may yield false positive or negative results as the enzyme reacts with abundantly available glucan molecules derived from yeast and plant sources [6]. Matrix-assisted laser desorption ionization/time of flight mass spectroscopy (MALDI-TOF MS) has also been used to identify bacterial endotoxins accurately, but this method is very expensive and time-consuming and requires specific expertise [7]. New methods that can rapidly and reliably detect bacterial endotoxins with a low limit of detection are highly desired in the fields of human health, environmental monitoring, and food safety.

In recent years, with the development of nanotechnology, biosensors based on novel nanostructures have been used to detect and identify trace amounts of endotoxins in human fluid samples using fluorescence, chemiluminescence, and electrical gradient applications [8–10]. Surface-enhanced Raman scattering (SERS) is considered to be one of the most sensitive analytical tools, with the potential to perform single-molecule detection under ambient conditions [11]. Recent ultra-sensitive SERS-based methods can detect single molecule reactions on a substrate surface [12,13]; however, Raman-based studies on the unique spectra of bacterial endotoxin are currently very limited. A recent study reported the in situ detection of a *Pseudomonas aeruginosa* endotoxin using a nanogold-sputtered cicada wing as a SERS chip; however, it used 4-mercaptobenzoic acid (4-MBA) and p-aminophenol (PAP) as SERS reporters rather than the intrinsic SERS spectra of the endotoxin itself [14]. In order to establish a highly sensitive SERS platform for intrinsic endotoxin detection, highly reproducible and practical substrates are essential. Our previous studies have demonstrated that silver nanorods (AgNRs) fabricated by oblique angle deposition can act as a highly sensitive and reproducible substrate with a SERS enhancement factor of $\sim 10^9$ and a batch-to-batch variation of $<10\%$ [15]. AgNR substrates have been applied in the detection of viruses, pathogenic bacteria such as *E. coli* O157:H7, *Salmonella typhimirium*, and *Staphylococcus aureus*, and toxins such as aflatoxins [16]. The utility of SERS from silver nanorods (AgNRs) for the rapid detection of bacterial pathogens via biomarkers has been demonstrated. We have reported that SERS can rapidly detect *Pseudomonas aeruginosa* pigment, pyocyanin, and pyoverdine in biological fluids with very high sensitivity and specificity when using a AgNR substrate [17].

Here, we report a proof-of-concept study in which SERS from AgNR is utilized to investigate the endotoxin structures of pathogenic bacteria. Highly purified and defined endotoxins from *N. meningitidis* and other pathogenic enteric bacteria are used to determine the unique SERS spectra. Utilizing principal component analysis (PCA), the unique SERS fingerprint spectra are able to accurately distinguish between meningococcal LPS structures and a collection of enteric LPS structures.

2. Materials and Methods

For endotoxin (LPS) sample preparation, the *Neisseria meningitidis* lipooligosaccharide (herein called LPS) and its truncated LPS structures (KDO_2-lipid A and unglycosylated lipid A) were isolated from *N. meningitidis* serogroup B and its isogenic mutants as described previously [2]. Of note, the *Neisseria meningitidis* endotoxin structure contains one side saccharide chain, hence referred to as lipooligosaccharide (LOS), which was used to distinguish between other lipopolysaccharide (LPS) structures. Purified LPSs from the enteric bacteria *E. coli*, *Vibrio cholerae*, *Salmonella typhimurium*, and *Salmonella minnesota* were obtained from Sigma (St. Louis, MO, USA) and further purified and quantified based on the lipid A content as previously described [4]. Briefly, residual membrane phospholipids were removed by the repeated extraction of the dried LPS samples with ethanol and water in a ratio of 9:1. The expected fatty acyl components of 3-OHC12:0, 3-OHC14:0, and C12:0 and the absence of membrane phospholipids were assessed by mass spectroscopy (GC-MS) (Dr. Russell Carlson, Complex Carbohydrate Research Center, University of Georgia, Athens, GA). LPS stock solutions were prepared in pyrogen-free water at a 10 nmole/mL concentration with extensive vortexing and sonication prior to each dilution as described previously [4]. Highly purified LPSs from *Rhizobium etli* strain CE3 and *Rhizobium niger* strain NGR (obtained as kind gifts from Dr. Russell Carlson, Complex Carbohydrate Research Center, University of Georgia, Athens, GA) were used at 10 nmole/mL concentrations. Lipid A is the major component of endotoxin structure responsible for biological activity. As such, all endotoxin structures used in this study were quantified based on their lipid A fatty acyl chain content rather than the total weight of molecule with and without saccharides chains, and stocks are made at 10 nmole/mL as described above [4].

Regarding silver nanorod (AgNRs) substrate fabrication, the AgNRs substrates used in this study were fabricated by an oblique angle deposition (OAD) technique that has been described previously [15,16,18]. Briefly, microscopic glass slides (BD, Portsmouth, NH) were cleaned with a piranha solution (80% sulfuric acid, 20% hydrogen peroxide, v/v) and rinsed with deionized (DI) water before being air-dried and loaded into a custom-made electron beam evaporation system. The glass slides were deposited with a 20-nm titanium layer and then a 200-nm silver film layer through evaporation at rates of ~0.2 nm/s and 0.3 nm/s, respectively. The slides were monitored in situ by a quartz crystal microbalance (QCM). To fabricate the AgNRs, the substrates were tilted to 86° with respect to the incident vapor, and silver was then deposited rate of ~0.3 nm/s until a thickness of 2000 nm was obtained. The morphology of the AgNR substrate has been reported previously [19–21]. According to a previous study, the AgNR substrate has a broad and strong absorption when the wavelength of light is >500 nm [19–21]. The uniformity and reproducibility of the AgNR substrate for SERS measurements have been reported to be smaller than 10% for the spot-to-spot variation and less than 15% for batch-to-batch variation [19].

Regarding SERS measurement and data analysis, 2 μL of the testing sample was dispensed directly on the AgNR substrate and vacuum-dried in the custom-made vacuum chamber to shorten the sample preparation time. The dried samples on the substrates were then measured by an Enwave ProRaman-L-785A2 Raman analyzer (Enwave Optronics, Irvine, CA, USA) with a 785 nm near-IR diode laser as the excitation source. SERS spectra were collected in a wavenumber range from 400 cm^{-1} to 1800 cm^{-1} for 10 to 30 s with a laser power on the sample of 100 mW. Spectra data were acquired from nine randomly selected spots on the AgNR substrate and showed good reproducibility (Appendix C). The original SERS spectra contain a broad fluorescence background, and thus all spectra baselines were first corrected and the average spectra from the nine different measurements per sample were presented as the final spectra results.

Data analysis was performed using version 9.0 of the Origin software package (OriginLab Corporation, Northampton, MA, USA). PCA was conducted with MATLAB 2000b (The MathWorks, Inc., Natick, MA, USA) using the PLS toolbox (Eigenvector Research, Inc.,

Wenatchee, WA, USA). Savitzky–Golay derivation, normalization, and the mean-center process were used to treat the raw spectra data prior to PCA analysis.

3. Results and Discussion

3.1. Meningococcal SERS Spectra

In order to understand the effect of a LPS contributing to its unique SERS spectra, a *Neisseria meningitidis* LPS (wild type, denoted as LPS), *Neisseria meningitidis* truncated LPS (KDO$_2$-lipid A), and *Neisseria meningitidis* unglycosylated lipid A (lipid A) at concentration of 10 nmole/mL were used to obtain SERS spectra. As shown in Figure 1a, the SERS spectrum of a LPS shows significant Raman shift peaks at $\Delta\nu$ = 989 cm^{-1} and $\Delta\nu$ = 1330 cm^{-1}, whereas the SERS spectra of KDO$_2$-lipid A has the same significant peaks as those observed from wild-type LPS spectra. Lipid A, the fatty acyl chain linked to the disaccharide backbone of LPS, is primarily responsible for the endotoxic activity in the human host, rather than the sugar chains in LPS [2]. When comparing the SERS signals from naked lipid A structure to those of the LPS with saccharide chains attached, the resulting SERS spectra suggest that the lipid A portion of the LPS could contribute significantly to its unique SERS signal. The SERS spectrum of KDO$_2$-lipid A shows a relatively strong peak at $\Delta\nu$ = 989 cm^{-1} and a weak peak at $\Delta\nu$ = 1330 cm^{-1} as compared to those for lipid A, which indicates that the characteristic peak at $\Delta\nu$ = 989 cm^{-1}, which may come from carbohydrates [22]. The peak at $\Delta\nu$ = 1330 cm^{-1} is typically assigned to δ(CH) in phospholipids, which is often used in the in vivo study of biological tissues [23]. The SERS spectra of lipid A have a relatively lower peak at $\Delta\nu$ = 989 cm^{-1} and a much higher peak intensity at $\Delta\nu$ = 1330 cm^{-1}, which further suggests that the peak at $\Delta\nu$ = 989 cm^{-1} is corresponding to the carbohydrates since KDO molecule is a sugar, and the signal from phospholipids is dominant in SERS spectra (Figure 1a). Similarly, the peak at $\Delta\nu$ = 1131 cm^{-1}, which was assigned to the mode of fatty acid, which also dominates the SERS spectra of lipid A [22,24]. Detailed SERS peak assignments are listed in Table 1.

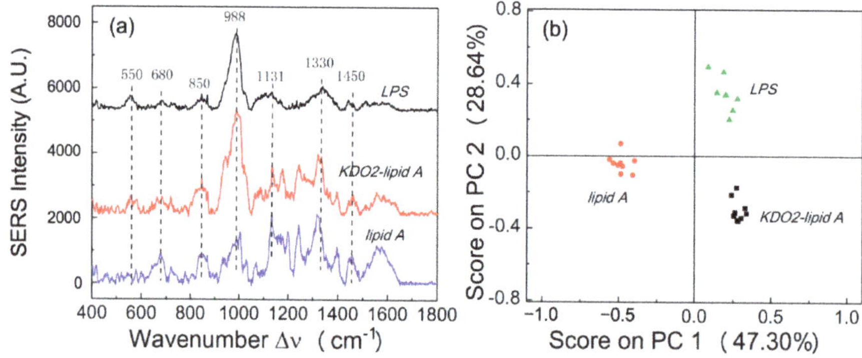

Figure 1. Meningococcal LPS Raman SERS spectra. (**a**) SERS spectra of LPS, KDO$_2$-lipid A, and lipid A obtained from *Neisseria meningitidis* endotoxin structures. (**b**) The PCA score plot of SERS spectra of LPS, KDO$_2$-lipid A, and lipid A.

The data suggest that the SERS spectra from different meningococcal LPS structures can be distinguished based on their detailed compositions. To better analyze the data, we have also performed the PCA analysis as shown in Figure 1b. The spectra of *Neisseria meningitidis* LPS, KDO$_2$-lipid A, and naked lipid A are well separated based on the score plots of PC1 (47.30%) and PC2 (28.64%). This study demonstrates the detection of highly purified meningococcal endotoxin structures and the differentiation of their SERS spectra in relation to the biological activity in a human host [2,4]. As previously reported, full-length LPSs and KDO$_2$-lipid A show potent immune stimulatory activity in human macrophages, whereas naked lipid A is a very weak endotoxin [2]. The biological activities of endotoxin structures are correlated with fulminant sepsis and disease outcomes [25,26]. Therefore, SERS can be used for the rapid detection and differentiation of a meningococcal endotoxin, which may serve as an indicator of *Neisseria meningitidis* infection.

Table 1. SERS peak assignment for *Neisseria meningitidis* LPS structures.

Observed SERS Shift $\Delta\nu$ (cm^{-1})	Vibrational Mode Assignment [22,27,28]
550	$\beta(CH_2)$ in ring
680	δ(C-O-C); fatty acid;
850	ν(C-O-C); saccharides (1,4 glycosidic link)
988	β(CH); carbohydrates
1131	ν(C-C); fatty acid
1330	δ(CH); phospholipid
1450	$\alpha(CH_3/CH_2)$, $\beta(CH_3/CH_2)$

β, bending; ν, stretching; δ, deformation.

3.2. Enteric LPS SERS Spectra

The SERS spectra of enteric LPSs at equal molar concentrations of 10 nmole/mL from *E. coli*, *S. typhimurium*, *S. minnesota*, *V. cholerae*, *R.* NGR, and *R.* CE3 are shown in Figure 2a. The spectra of *S. typhimurium* and *S. minnesota*, which are from the same species (*Salmonella*), show similar peaks at $\Delta\nu$ = 543 and 981 cm^{-1}, which can be signed to the saccharide and lipid [22,24,27,29]. Similarly, the SERS spectra of *R.* CE3 and *R.* NGR also show the same peaks at $\Delta\nu$ = 552, 981, and 1330 cm^{-1} (Figure 2a), which represents cholesterol and lipids [22,24,27–30]. The spectra of *E. coli* and *V. cholerae* are entirely different from the spectra of the *Salmonella* LPS and *Rhizobium* LPS, which indicates that unique the SERS spectra of LPS structures could be used to differentiate bacteria species (Figure 2a). Detailed SERS peak assignment for enteric and *Rhizobium* LPS are shown in Table 2. Note that most of the spectral peaks of these LPSs are from lipids and carbohydrates, which are the major components of LPS. The 3D-score plot of PC 1 (42.20%), PC 2 (24.82%), and PC 3 (17.76%) for the enteric and *Rhizobium* LPS based on the SERS spectra are shown in Figure 2b. All the spectra from the same LPS are grouped together well and are evidently separated from one another. Different bacteria express unique LPS structures that do not change when expressed in viable cells or shed from the membrane. Therefore, unique LPS SERS can be used to distinguish various bacteria strains.

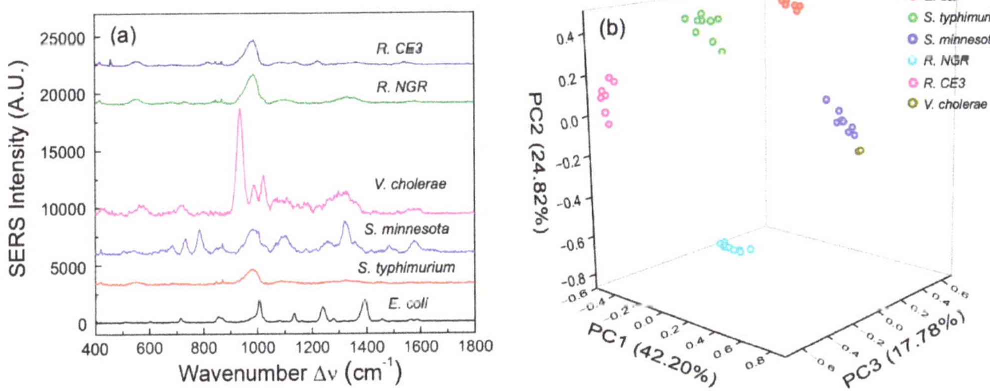

Figure 2. Different enteric LPS Raman SERS spectra. (**a**) SERS spectra of different enteric LPSs obtained from *E. coli*, *S. typhimurium*, *S. minnesota*, *V. cholerae*, *R.* CE3, and *R.* NGR. (**b**) PCA score plot of six enteric LPSs based on their SERS spectra.

Table 2. SERS peak assignment for enteric LPS structures.

Observed SERS Shift $\Delta\nu$ (cm^{-1})	Vibrational Mode Assignment [22,27,30–32]
543	δ(C-O-C) in glycosidic ring
552	β(CH_2) in ring
715	ν(C-N)
733	β(C-O-C) in carbohydrates
787	ν(C-O) in ring
855	δ(C-O-C)
981	β(CH) in lipid
1025	ν(CO) in carbohydrates
1087	ν(C-O) in lipid
1131	ν(C-C) in fatty acid
1309	τ(CH_3/CH_2) in lipid
1330	δ(CH) in phospholipid

β, bending; δ, deformation; τ, twisting; ν, stretching.

The SERS spectral peak locations are usually referred to as fingerprint spectra. Different molecules have different fingerprint spectra, which can be used to differentiate between different molecular structures [16]. It is known that LPSs from different bacteria have different molecular structures and bonds [4]. Therefore, our data analysis focused on the peak location difference rather than intensity or peak shape. Since all spectra were normalized, the peak intensities were less important when comparing various LPS structures. In our study, the PCA method was employed to depict the differences between peak intensity, peak position, relative intensity and spectral shape (Figure 2B). Although minor structural differences between bacterial LPSs were found, their biological activities were different and correlated to their effect on human host response [4]. For example, *Rhizobium* species that contain extended length fatty acyl chains do not exert immune stimulatory activity on human macrophages, which is in contrast to other LPSs like *E. coli*, *Salmonella* and *Vibrio* with fatty acyl chain lengths of C12 and C14. Moreover, *Salmonella* and *E. coli* LPS structures vary in their biological activity and differential induction of human TLR4 signaling pathways, consequently modulating the outcome of host response [4]. Taken together, the data provide a proof-of-concept for the utility of SERS in rapid detection of endotoxin and in distinguishing endotoxin structures from bacterial pathogens.

4. Conclusions

In this work, we have shown that the LPS, KDO_2-lipid A, and lipid A of *Neisseria meningitidis* endotoxin structures, as well as the enteric LPSs from *E. coli*, *S. typhimurium*, *S. minnesota*, *V. cholerae*, *R.* CE3, and *R.* NGR, exhibit distinguishable fingerprint SERS spectra that can be separated well using a conventional PCA method. The difference in SERS spectra originates from the carbohydrates and phospholipids of different LPS structures. All the SERS measurements were carried at a low concentration (10 nmole/mL) in accordance with the relevant pathophysiological concentration in clinical infection, which means that the SERS technique has a great potential to detect and differentiate physiologically relevant enteric LPSs from biological fluids. In addition, the spectral differences of the different LPSs demonstrate that LPSs can be used as biomarkers for detecting bacteria, especially in terms of SERS-based detection, which could potentially lead to new ways to sense bacteria.

Author Contributions: Conceptualization, Y.Z. and S.M.Z.; methodology, X.W., Y.Z., S.M.Z.; validation, X.W., Y.Z., S.M.Z.; formal analysis, X.W., Y.Z., S.M.Z.; investigation, X.W., Y.Z., S.M.Z.; data curation, X.W., Y.Z., S.M.Z.; writing—original draft preparation, X.W., S.M.Z.; writing—review and editing, X.W., Y.Z., S.M.Z.; supervision, Y.Z., S.M.Z.; project administration, Y.Z., S.M.Z.; funding acquisition, Y.Z., S.M.Z., Y.Z., S.M.Z. conceived the study; Y.Z., S.M.Z. designed the experiments, X.W. performed the experiments; X.W., Y.Z., S.M.Z. analyzed the data; X.W., Y.Z., S.M.Z. wrote the manuscript. All authors have read and agreed to the published version of the manuscript.

Funding: This research was funded in part by National Institute of Health grant R21-AI096364 and National Science Foundation under contract number grant CBET-1064228. This study was made possible in part by NPRP grant NPRP12S-0224-190144 from the Qatar National Research Fund (a member of the Qatar Foundation). The findings achieved herein are solely the responsibility of the authors. https://www.qnrf.org/en-us (accessed on 4 June 2021). https://grants.nih.gov/grants (accessed on 4 June 2021).

Institutional Review Board Statement: Not applicable.

Informed Consent Statement: Not applicable.

Data Availability Statement: Available upon request.

Acknowledgments: The authors also thank Funing Chen's technical help in generating some SERS spectra measurements.

Conflicts of Interest: The authors declare no conflict of interest.

Appendix A

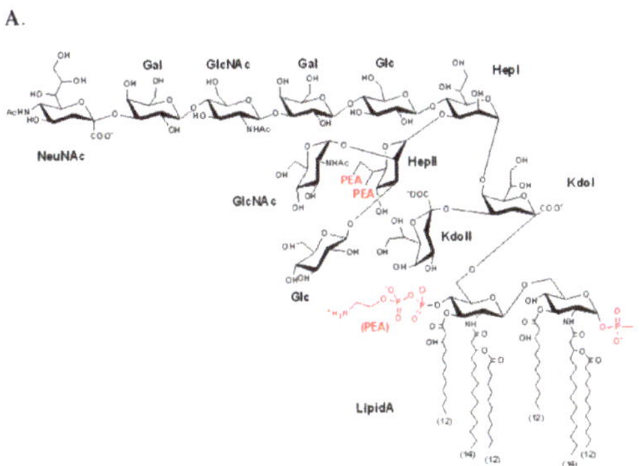

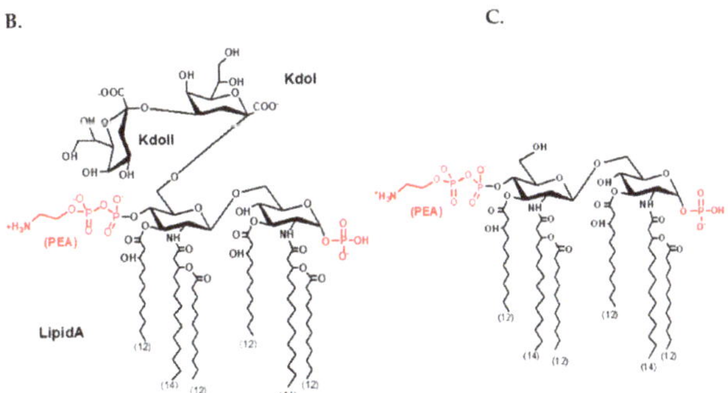

Figure A1. Schematic representation of meningococcal LPS structures. (**A**) *N. meningitidis* strain NMB wild-type LPS structure; (**B**) minimally glycosylated meningococcal KDO_2-lipid A structure; (**C**) unglycosylated meningococcal lipid A structure.

Appendix B

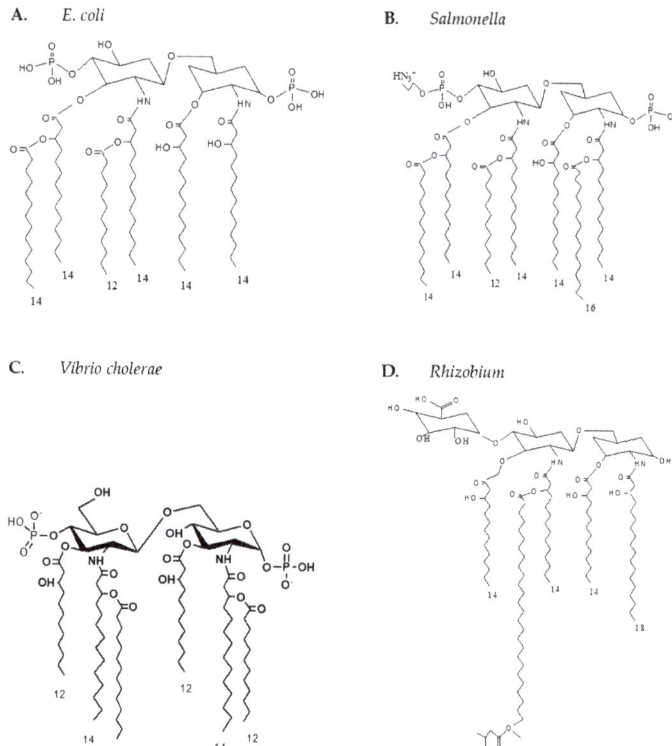

Figure A2. Schematic representation of different enteric lipid A structures. (**A**): *E. coli;* (**B**): *Salmonella;* (**C**): *Vibrio cholerae;* (**D**): *Rhizobium.*

Appendix C

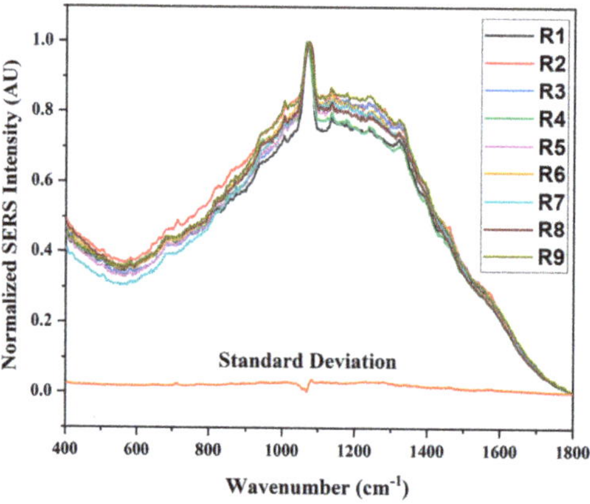

Figure A3. Reproducibility of LPS spectra on AgNR. Nine spectra (R1 to R9) are shown for the same LPS as measured at different locations on the AgNR substrate.

References

1. Park, B.S.; Song, D.H.; Kim, H.M.; Choi, B.-S.; Lee, H.; Lee, J.-O. The structural basis of lipopolysaccharide recognition by the TLR4-MD-2 complex. *Nature* **2009**, *458*, 1191–1195. [CrossRef]
2. Zughaier, S.M.; Tzeng, Y.-L.; Zimmer, S.M.; Datta, A.; Carlson, R.W.; Stephens, D.S. *Neisseria meningitidis* lipooligosaccharide structure-dependent activation of the macrophage CD14/Toll-like receptor 4 pathway. *Infect. Immun.* **2004**, *72*, 371–380. [CrossRef] [PubMed]
3. Brandtzaeg, P.; Ovstebo, R.; Kierulf, P. Bacteremia and compartmentalization of LPS in meningococcal disease. *Prog. Clin. Biol. Res.* **1995**, *392*, 219–233. [PubMed]
4. Zughaier, S.M.; Zimmer, S.M.; Datta, A.; Carlson, R.W.; Stephens, D.S. Differential induction of the toll-like receptor 4-MyD88-dependent and -independent signaling pathways by endotoxins. *Infect. Immun.* **2005**, *73*, 2940–2950. [CrossRef] [PubMed]
5. Forsberg, L.S.; Carlson, R.W. The structures of the lipopolysaccharides from *Rhizobium etli* strains CE358 and CE359. The complete structure of the core region of *R. etli* lipopolysaccharides. *J. Biol. Chem.* **1998**, *273*, 2747–2757. [CrossRef]
6. Iwanaga, S. Biochemical principle of Limulus test for detecting bacterial endotoxins. *Proc. Jpn. Acad. Ser. B Phys. Biol. Sci.* **2007**, *83*, 110–119. [CrossRef]
7. Larrouy-Maumus, G.; Clements, A.; Filloux, A.; McCarthy, R.; Mostowy, S. Direct detection of lipid A on intact Gram-negative bacteria by MALDI-TOF mass spectrometry. *J. Microbiol. Methods* **2016**, *120*, 68–71. [CrossRef] [PubMed]
8. Wu, Y.; Deng, P.; Tian, Y.; Feng, J.; Xiao, J.; Li, J.; Liu, J.; Li, G.; He, Q. Simultaneous and sensitive determination of ascorbic acid, dopamine and uric acid via an electrochemical sensor based on PVP-graphene composite. *J. Nanobiotechnol.* **2020**, *18*, 112. [CrossRef]
9. Kim, S.-E.; Su, W.; Cho, M.; Lee, Y.; Choe, W.-S. Harnessing aptamers for electrochemical detection of endotoxin. *Anal. Biochem.* **2012**, *424*, 12–20. [CrossRef]
10. Seth, R.; Ribeiro, M.; Romaschin, A.; Scott, J.A.; Manno, M.; Scott, J.A.; Liss, G.M.; Tarlo, S.M. Occupational endotoxin exposure and a novel luminol-enhanced chemiluminescence assay of nasal lavage neutrophil activation. *J. Allergy Clin. Immunol.* **2011**, *127*, 272–275. [CrossRef]
11. Kneipp, J.; Kneipp, H.; Kneipp, K. SERS—A single-molecule and nanoscale tool for bioanalytics. *Chem. Soc. Rev.* **2008**, *37*, 1052–1060. [CrossRef]
12. Choi, H.-K.; Lee, K.S.; Shin, H.-H.; Koo, J.-J.; Yeon, G.J.; Kim, Z.H. Single-Molecule Surface-Enhanced Raman Scattering as a Probe of Single-Molecule Surface Reactions: Promises and Current Challenges. *Acc. Chem. Res.* **2019**, *52*, 3008–3017. [CrossRef]
13. Fang, W.; Jia, S.; Chao, J.; Wang, L.; Duan, X.; Liu, H.; Li, Q.; Zuo, X.; Wang, L.; Wang, L.; et al. Quantizing single-molecule surface-enhanced Raman scattering with DNA origami metamolecules. *Sci. Adv.* **2019**, *5*, eaau4506. [CrossRef]
14. Xiang, S.; Ge, C.; Li, S.; Chen, L.; Wang, L.; Xu, Y. In Situ Detection of Endotoxin in Bacteriostatic Process by SERS Chip Integrated Array Microchambers within Bioscaffold Nanostructures and SERS Tags. *ACS Appl. Mater. Interfaces* **2020**, *12*, 28985–28992. [CrossRef]
15. Driskell, J.D.; Shanmukh, S.; Liu, Y.; Chaney, S.; Hennigan, S.; Jones, L.; Krause, D.; Tripp, R.A.; Zhao, Y.-P.; Dluhy, R.A. Novel Nanoarray SERS Substrates Used for High Sensitivity Virus Biosensing and Classification. *Nanosci. Nanotechnol. Chem. Biol. Def.* **2009**, *1016*, 99–114.
16. Wu, X.; Chen, J.; Park, B.; Huang, Y.W.; Zhao, Y. The Use of Silver Nanorod Array-Based Surface-Enhanced Raman Scattering Sensor for Food Safety Applications, in Advances in Applied Nanotechnology for Agriculture, in Advances in Applied Nanotechnology for Agriculture. *Am. Chem. Soc.* **2013**, *1143*, 85–108.
17. Wu, X.; Chen, J.; Li, X.; Zhao, Y.; Zughaier, S.M. Culture-free diagnostics of *Pseudomonas aeruginosa* infection by silver nanorod array based SERS from clinical sputum samples. *Nanomedicine* **2014**, *10*, 1863–1870. [CrossRef]
18. Liu, Y.J.; Chu, H.Y.; Zhao, Y.P. Silver Nanorod Array Substrates Fabricated by Oblique Angle Deposition: Morphological, Optical, and SERS Characterizations. *J. Phys. Chem. C* **2010**, *114*, 8176–8183. [CrossRef]
19. Driskell, J.D.; Shanmukh, S.; Liu, Y.; Chaney, S.B.; Tang, X.J.; Zhao, Y.P.; Dluhy, A.R. The Use of Aligned Silver Nanorod Arrays Prepared by Oblique Angle Deposition as Surface Enhanced Raman Scattering Substrates. *J. Phys. Chem. C* **2008**, *112*, 895–901. [CrossRef]
20. Zhao, Y.P.; Chaney, S.B.; Zhang, Z.Y. Absorbance spectra of aligned Ag nanorod arrays prepared by oblique angle deposition. *J. Appl. Phys.* **2006**, *100*, 063527. [CrossRef]
21. Abell, J.; Driskell, J.; Dluhy, R.; Tripp, R.; Zhao, Y.-P. Fabrication and characterization of a multiwell array SERS chip with biological applications. *Biosens. Bioelectron.* **2009**, *24*, 3663–3670. [CrossRef] [PubMed]
22. Czamara, K.; Majzner, K.; Pacia, M.Z.; Kochan, K.; Kaczor, A.; Baranska, M. Raman spectroscopy of lipids: A review. *J. Raman Spectrosc.* **2015**, *46*, 4–20. [CrossRef]
23. Utzinger, U.; Heintzelman, D.L.; Mahadevan-Jansen, A.; Malpica, A.; Follen, M.; Richards-Kortum, R. Near-infrared Raman spectroscopy for in vivo detection of cervical precancers. *Appl. Spectrosc.* **2001**, *55*, 955–959. [CrossRef]
24. Krafft, C.; Neudert, L.; Simat, T.; Salzer, R. Near infrared Raman spectra of human brain lipids. *Spectrochim. Acta Part A Mol. Biomol. Spectrosc.* **2005**, *61*, 1529–1535. [CrossRef] [PubMed]
25. Brandtzaeg, P.; Bryn, K.; Kierulf, P.; Ovstebø, R.; Namork, E.; Aase, B.; Jantzen, E. Meningococcal endotoxin in lethal septic shock plasma studied by gas chromatography, mass-spectrometry, ultracentrifugation, and electron microscopy. *J. Clin. Investig.* **1992**, *89*, 816–823. [CrossRef] [PubMed]

26. Zughaier, S.M.; Lindner, B.; Howe, J.; Garidel, P.; Koch, M.H.; Brandenburg, K.; Stephens, D.S. Physicochemical characterization and biological activity of lipooligosaccharides and lipid A from *Neisseria meningitidis*. *J. Endotoxin Res.* **2007**, *13*, 343–357. [CrossRef]
27. Wiercigroch, E.; Szafraniec, E.; Czamara, K.; Pacia, M.Z.; Majzner, K.; Kochan, K.; Kaczor, A.; Baranska, M.; Malek, K. Raman and infrared spectroscopy of carbohydrates: A review. *Spectrochim. Acta Part A Mol. Biomol. Spectrosc.* **2017**, *185*, 317–335. [CrossRef]
28. Ruggiero, L.; Sodo, A.; Bruni, F.; Ricci, M.A. Hydration of monosaccharides studied by Raman scattering. *J. Raman Spectrosc.* **2018**, *49*, 1066–1075. [CrossRef]
29. Notingher, I.; Green, C.; Dyer, C.; Perkins, E.; Hopkins, N.; Lindsay, C.; Hench, L.L. Discrimination between ricin and sulphur mustard toxicity in vitro using Raman spectroscopy. *J. R. Soc. Interface* **2004**, *1*, 79–90. [CrossRef]
30. Malini, R.; Venkatakrishna, K.; Kurien, J.; Pai, K.M.; Rao, L.; Kartha, V.B.; Krishna, C.M. Discrimination of normal, inflammatory, premalignant, and malignant oral tissue: A Raman spectroscopy study. *Biopolymers* **2006**, *81*, 179–193. [CrossRef] [PubMed]
31. Lu, X.; Al-Qadiri, H.M.; Lin, M.; Rasco, B.A. Application of Mid-infrared and Raman Spectroscopy to the Study of Bacteria. *Food Bioprocess Technol.* **2011**, *4*, 919–935. [CrossRef]
32. Delgado-Coello, B.; Montalvan-Sorrosa, D.; Cruz-Rangel, A.; Sosa-Garrocho, M.; Hernández-Téllez, B.; Macías-Silva, M.; Castillo, R.; Mas-Oliva, J. Label-free surface-enhanced Raman spectroscopy of lipid-rafts from hepatocyte plasma membranes. *J. Raman Spectrosc.* **2017**, *48*, 659–667. [CrossRef]

Communication

Self-Powered Biosensor for Specifically Detecting Creatinine in Real Time Based on the Piezo-Enzymatic-Reaction Effect of Enzyme-Modified ZnO Nanowires

Meng Wang, Guangting Zi, Jiajun Liu, Yutong Song, Xishan Zhao, Qi Wang * and Tianming Zhao *

College of Sciences, Northeastern University, Shenyang 110819, China; 2000176@stu.neu.edu.cn (M.W.); 20191534@stu.neu.edu.cn (G.Z.); 2000165@stu.neu.edu.cn (J.L.); 2000172@stu.neu.edu.cn (Y.S.); 2000189@stu.neu.edu.cn (X.Z.)

* Correspondence: wangqi@mail.neu.edu.cn (Q.W.); zhaotm@stumail.neu.edu.cn (T.Z.)

Abstract: Creatinine has become an important indicator for the early detection of uremia. However, due to the disadvantages of external power supply and large volume, some commercial devices for detecting creatinine concentration have lost a lot of popularity in everyday life. This paper describes the development of a self-powered biosensor for detecting creatinine in sweat. The biosensor can detect human creatinine levels in real time without the need for an external power source, providing information about the body's overall health. The piezoelectric output voltage of creatininase/creatinase/sarcosine oxidase-modified ZnO nanowires (NWs) is significantly dependent on the creatinine concentration due to the coupling effect of the piezoelectric effect and enzymatic reaction (piezo-enzymatic-reaction effect), which can be regarded as both electrical energy and biosensing signal. Our results can be used for the detection of creatinine levels in the human body and have great potential in the prediction of related diseases.

Keywords: creatinine; ZnO nanowires; piezo-enzymatic-reaction effect; self-powered biosensor

Citation: Wang, M.; Zi, G.; Liu, J.; Song, Y.; Zhao, X.; Wang, Q.; Zhao, T. Self-Powered Biosensor for Specifically Detecting Creatinine in Real Time Based on the Piezo-Enzymatic-Reaction Effect of Enzyme-Modified ZnO Nanowires. *Biosensors* **2021**, *11*, 342. https://doi.org/10.3390/bios11090342

Received: 24 August 2021
Accepted: 14 September 2021
Published: 16 September 2021

1. Introduction

Following the fast development of the social economy, people's demands for a better life are not just based on food or warm clothes but also on good health [1–3]. The maintenance of a healthy lifestyle is inextricably linked to the monitoring of various body indexes [4,5], with creatinine serving as an important index for the early detection of uremia [6]. Creatinine [6–8], one of the most important metabolic health indicators [9–11], can indicate abnormal renal function as well as other serious diseases [12–14]. However, almost all traditional creatinine sensors require external power supplies and large volumes [15,16], making the sensor inconvenient to transport and posing a risk due to the frequent charging/discharging process [17]. Those features restrict the applications of those devices in many areas such that they cannot be used for daily health detection.

To solve the common problem of portability in capacitor/resistance-based sensors [18], many scientists have conducted in-depth research [19]. Using piezoelectric materials or triboelectric structures to realize self-powered sensors has been a popular solution in recent years. The self-powered nanogenerators can convert external mechanical energy into electrical energy [20,21], and the output is easily affected by the external environment [22] by doping noble metal elements or modifying functional materials [23–31], which can be viewed as a sensing signal. Traditional triboelectric nanogenerators based on a vertical contraction–separation structure or a contact–slide structure are not flexible enough due to their unique structure [32,33]. Furthermore, common piezoelectric semiconductors are incapable of detecting biological materials [34]. Therefore, it is essential to design a self-powered sensor [35] for specifically detecting creatinine.

In this paper, a self-powered biosensor for specifically detecting creatinine in sweat is presented. Based on the coupling effect of the piezoelectric effect and enzymatic reaction [36–38] (piezo-enzymatic-reaction effect) of the creatininase/creatinase/sarcosine oxidase-modified ZnO NWs, the biosensor can actively output electrical signals, which can be seen as biological signals. In addition, the piezoelectric output voltage is significantly dependent on the creatinine concentration, and therefore can reflect the concentration of creatinine. Hence, the whole process does not need an external power supply. Our research results represent a great contribution to the detection of creatinine levels in the human body.

2. Materials and Methods

2.1. Materials

Zinc acetate ($Zn(CH_3COO)_2{\cdot}2H_2O$), zinc nitrate hexahydrate ($Zn(NO_3)_2{\cdot}6H_2O$), creatininase, creatinase, and sarcosine oxidase were purchased from Macklin Inc. (Shanghai, China) Ammonia solution ($NH_3{\cdot}H_2O$), creatinine, and phosphate-buffered saline (PBS) solution were provided from Sinopharm Chemical Reagent Co. Ltd. (Shanghai, China) Ti film (200 μm) and Kapton film (100 μm) were purchased from Taobao (Hangzhou, China). Photoresist and developing solutions were purchased from Suzhou Ruihong Electronic Chemicals Co. Ltd (Suzhou, China). All chemicals were analytically pure and utilized without further purification.

2.2. Synthesis of ZnO NWs

The ZnO NWs were synthesized by a simple seed-assisted hydrothermal method. First, a ZnO seed layer was grown on Ti foil. A precleaned Ti film was immersed in a $Zn(CH_3COO)_2{\cdot}2H_2O$ (10 mM in ethanol) solution for several seconds and then dried under nitrogen gas. The Ti film with $Zn(CH_3COO)_2$ nanoparticles was then annealed at 350 °C for 20 min to form the ZnO seed layer. Second, the ZnO NWs were synthesized on the ZnO seeds. The Ti film with the ZnO seed layer was immersed into a solution containing 0.82 g of ($Zn(NO_3)_2{\cdot}6H_2O$) and 2.5 mL of $NH_3{\cdot}H_2O$ in 38 mL of deionized water. After 24 h at 83 °C, the Ti film with the ZnO NWs was cleaned with deionized water and ethanol several times; the Ti foil was recycled for the next use after the transfer-printing process to further lower the fabrication cost.

2.3. Device Fabrication

The fabrication process of the device is mainly composed of photolithography, electron-beam evaporation, and culture process. The experimental process is described below in detail. First, the ZnO NWs of the Ti film were transferred to a Kapton film with a knife. Then, a ~2 μm photoresistor was spin-coated on the Kapton film at 2000 rpm for 70 s. Using a lithography process, a predesigned pattern was formed. Next, the Ti electrodes were formed by an electronic beam evaporation process. The thickness of the Ti electrodes was set at 200 nm. After removing the residual photoresist, the device was dried overnight with nitrogen gas. Finally, creatininase (10 u/mL), creatininase (10 u/mL), and sarcosine oxidase (10 u/mL) were added dropwise on the device and an incubation procedure was conducted in a biosafety cabinet for 2 h. For a better test outcome, the creatininase, creatinase, and sarcosine oxidase were dissolved in a PBS solution (pH ~7.4) at 20 °C. The self-powered creatinine biosensor was stored at 4 °C.

2.4. Characterization and Measurement

A scanning electron microscope (SEM; JEOL JSM-6700 F) equipped with an energy dispersive X-ray spectrometer (EDS) was used to characterize the devices' morphology. A low-noise preamplifier was used to measure the performance of the devices (Model SR560, Stanford Research Systems). A programmable system with a stepping motor and a sliding rail provided the force and working frequency applied to the devices. All experiments were carried out at a temperature of 25 °C and relative humidity of 40%.

3. Results

In recent years, creatinine has increasingly been tested in daily life, as an important biomarker to measure human health [39,40]. However, the majority of commercial instruments require an external power source, which make them bulky and unwieldy, stifling further development. In this study, a self-powered creatinine biosensor based on enzyme-modified ZnO NWs is demonstrated. Combining the piezoelectric effect of ZnO NWs with enzyme reaction (piezo-enzymatic-reaction effect), the device can actively output electrical pulses by harvesting tiny mechanical energy, which is significantly dependent on creatinine concentration. With increasing creatinine concentration, the piezoelectric output voltage decreases. Furthermore, because the piezoelectric output voltage can be regarded as both a power source and a biosignal, the entire process does not require any power supply.

Figure 1a shows the concept of a self-powered creatinine biosensor and Figure 1b shows an optical photograph of the self-powered creatinine biosensor. The predesigned pattern includes many interdigital electrode pairs. The distance between electrodes is 4 μm and the distance between each electrode pair is 20 μm to prevent electrical short circuit because the length of the ZnO NWs is ~7 μm (greater than 4 μm, less than 20 μm). Figure 1c shows the 45° view SEM of the ZnO NWs. It can be observed that the ZnO NWs are orderly grown on the Ti film and the diameter of the ZnO NWs is ~200 nm. In Figure 1d, a side view SEM of the ZnO NWs shows that the average length is ~7 μm. The length ensures the ZnO NWs cannot cross electrode pairs (less than 20 μm). Figure 1e shows a single ZnO NW bridge and electrode pair before removing the photoresist. The distance between an electrode pair is 4 μm. Figure 1f shows a single ZnO NW bridge and electrode pair after e-beam evaporating the Ti electrodes. It can be seen that the distance between each electrode pair is 20 μm. Figure 1g shows the X-ray diffraction pattern of the ZnO NWs. The triangle represents the characteristic peak of ZnO and the circle represents the characteristic peak of Ti. The peaks around 31.8°, 34.4°, 36.2°, 47.5°, 56.6°, 62.9°, 69.1°, 72.6°, 76.9°, 81.4°, and 89.6° correspond to the (100), (002), (101), (102), (110), (103), (200), (112), (201), (004), and (202) (PDF#89-0511), respectively. In addition, the peaks around 38.4°, 44.6°, and 64.9° can be attributed to the Ti substrate. Figure 1h shows the EDS spectrum of the ZnO NWs. The peaks of the Zn and O elements contribute to the ZnO NWs and the peaks of Ti contribute to the Ti film.

Figure 2 shows the fabrication of the self-powered creatinine biosensor. The process includes synthesis, transfer printing, photolithography, electron-beam evaporation of the Ti electrode, and enzyme culture of the ZnO NWs. First, during the hydrothermal synthesis of the ZnO NWs [36,37,41], the length of the ZnO NWs was adjusted by controlling the pH value of the solution and the synthesis time during the whole process. To ensure that the ZnO NWs can have the same orientation on the Kapton substrate, the vertical ZnO NWs need to be gently scraped with a knife for the transfer process. Then, the creatinine detector was prepared by photolithography, electron-beam evaporation, and enzyme modification on the prepared material. The specific process can be found in the Materials and Methods.

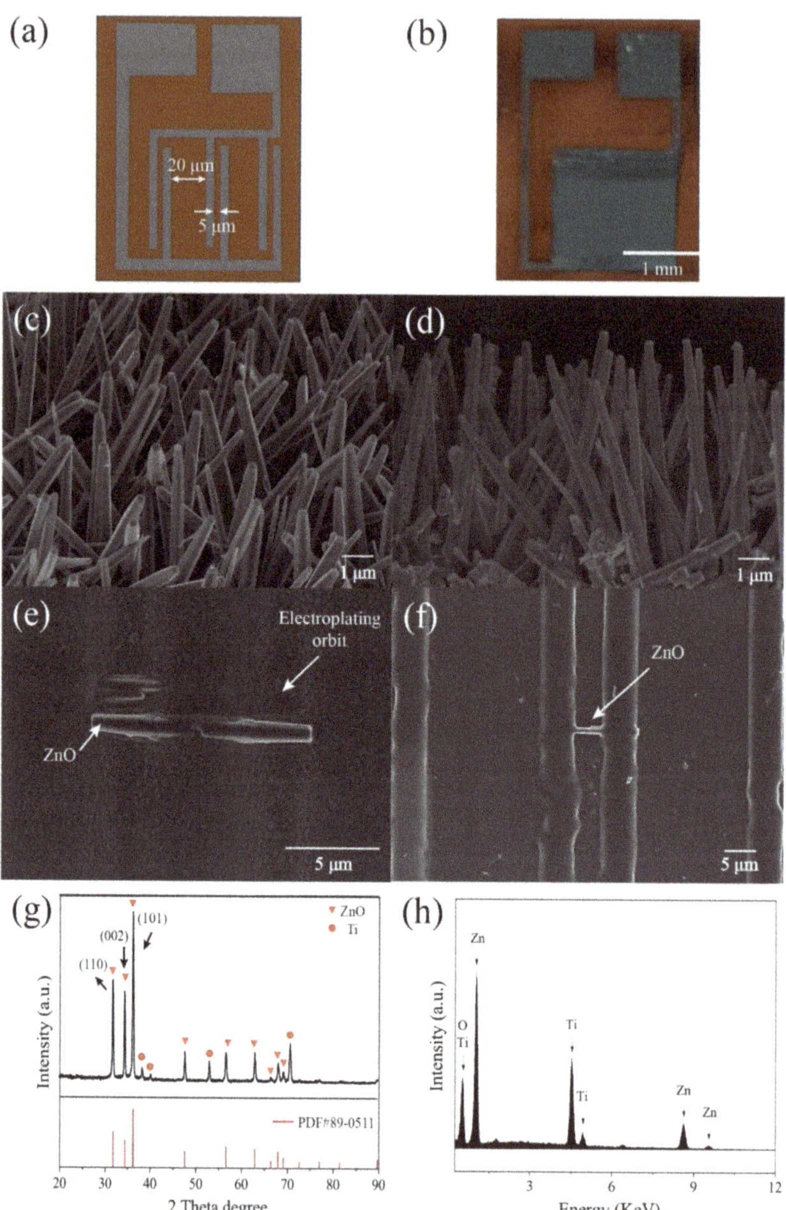

Figure 1. (**a**) Concept of self-powered creatinine biosensor. (**b**) Optical photograph of self-powered creatinine biosensor. (**c**) 45° view SEM of the ZnO NWs. (**d**) Side view SEM of the ZnO NWs. (**e**) A single ZnO NW bridge and electrode pair before removing the photoresist. (**f**) A single ZnO NW bridge and electrode pair after e-beam evaporating the Ti electrodes. (**g**) X-ray diffraction pattern of the ZnO NWs. (**h**) Energy dispersive X-ray spectrum of the ZnO NWs.

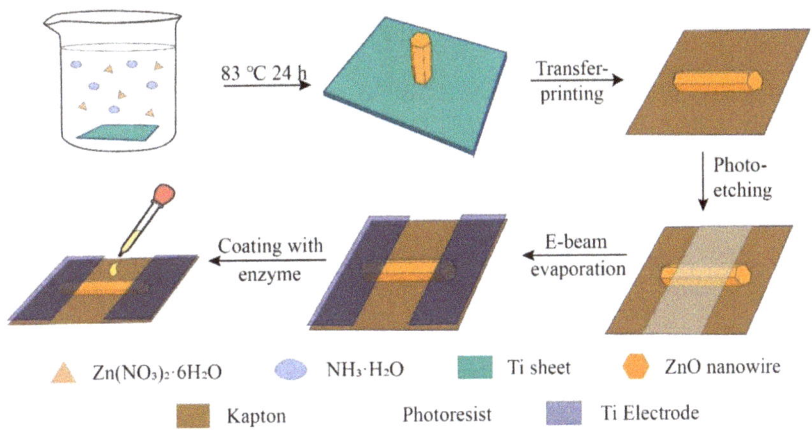

Figure 2. Fabrication of the self-powered creatinine biosensor.

The piezoelectric properties of the device were measured as shown in Figure 3. Figure 3a shows the schematic diagram of the measurement system. The device is put in a Petri dish filled with a PBS solution to mimic the sweat environment. To apply the working force and frequency, a programmable system with a stepping motor and a sliding rail is used. The device's electrodes are linked to the SR560 preamplifier (Model SR560, Stanford Research Systems), which collects piezoelectric signals and feeds them into a computer to monitor real-time voltage changes. A computer collects the piezoelectric output voltage. Figure 3b depicts the relationship between applied force and piezoelectric output voltage at 1 Hz and 0°, with the applied force shown in the inset. When the applied forces are 19, 22, and 25 N, the piezoelectric output voltages of the device are 0.24, 0.36, and 0.56 V, respectively. With the increase in pressure, the voltage value significantly increases. In addition, the linear relationship between the force and piezoelectric voltage output is:

$$y = -0.91083 + 0.05933 \times F, \quad (1)$$

where y represents the piezoelectric output voltage (in V) and F represents the applied force (in N). The square of the correlation coefficient is 0.93762, showing a great fit. Figure 3c shows the relationship between the bending angles and piezoelectric output voltage at 22 N and 1 Hz; the bending angles are defined as shown in the inset of Figure 3c. At bending angles of 30°, 45°, and 60°, the piezoelectric output voltages of the device are 0.14 V, 0.31 V, and 0.57 V, respectively. With the increase in bending angles, the piezoelectric output voltage significantly increases. In addition, the linear relationship between the angles and piezoelectric voltage output is:

$$y = -0.33852 + 0.01523 \times \theta, \quad (2)$$

where y represents the piezoelectric output voltage (in V) and θ represents the bending angle (in °). The square of the correlation coefficient is 0.97749, showing a great fit. Figure 3d shows the piezoelectric output voltage against different working frequencies at 22 N. The frequency has little effect on the piezoelectric output voltage. At 2, 1, and 0.5 Hz, the piezoelectric output voltage is almost stable (~0.3 V).

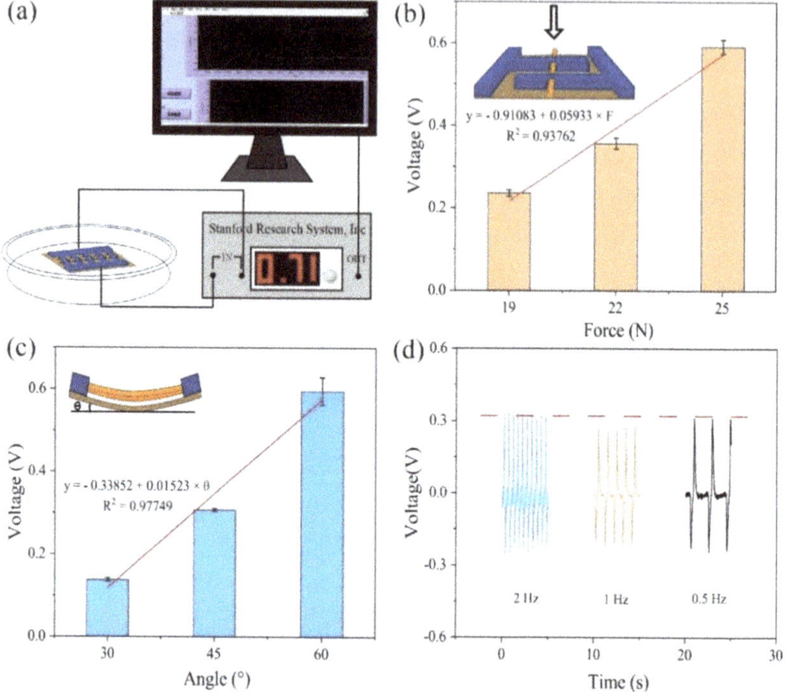

Figure 3. (**a**) Schematic diagram of the measurement system. (**b**) Relationship between applied force and piezoelectric output voltage at 1 Hz; the inset shows the applied force. (**c**) Relationship between the bending angles and piezoelectric output voltage at 22 N and 1 Hz; the bending angles are defined as shown in the inset. (**d**) Piezoelectric output voltage against different working frequencies at 22 N.

The sensing performances for detecting creatinine concentration are shown in Figure 4. Figure 4a shows the measurement process. To simulate the sweat environment, the self-powered biosensor is placed in a Petri dish filled with a PBS solution, and the solution containing different concentrations of creatinine is added to the Petri dish dropwise. The relationship between creatinine concentration and piezoelectric output voltage is depicted in Figure 4b. When the creatinine concentrations are 1×10^{-5}, 1×10^{-4}, 1×10^{-3}, 1×10^{-2}, and 1×10^{-1} mM, the piezoelectric outputs are 0.65, 0.41, 0.28, 0.21, and 0.12 V, respectively. With the increase in creatinine concentration, the piezoelectric output significantly decreases. Figure 5b shows that the detection range of creatinine concentration of the device is 1×10^{-5}–1×10^{-1} mM. For further analysis, the data between the log of concentration and piezoelectric output voltage is fitted by linear regression. The fit line is:

$$y = 0.0885 - 0.0229 \times \lg C, \quad (3)$$

where y represents piezoelectric output voltage (in V) and C represents creatinine concentration (in mM). The self-powered creatinine sensor has a sensitivity of 0.0229 V/mM. The sensing performance was tested at a concentration of 0.1 mM creatinine against various conditions (applied force, bending angle, and working frequency) shown in Figure 4c–e. Figure 4c shows the influence of creatinine concentration and force on piezoelectric output. When the creatinine concentration is 0 or 1 mM, 19, 22, and 25 N forces are applied to the material. As the creatinine concentration increases and the applied force decreases, the piezoelectric output significantly decreases. Similar results can be observed, which is contributed to the piezo-enzymatic-reaction effect. Figure 4f,g show the comparison of piezoelectric output variation trends under different applied forces, bending angles, and force frequencies before and after adding creatinine. Though the piezoelectric output voltage increases with the increasing applied force and bending angles before adding creatinine, the biosensing output after adding creatinine also increases with the increasing applied force and bending angles, which implies that the biosensing performance has little influence on applied force and bending angles (Figure 4f,g). Similar results can be seen in Figure 4h.

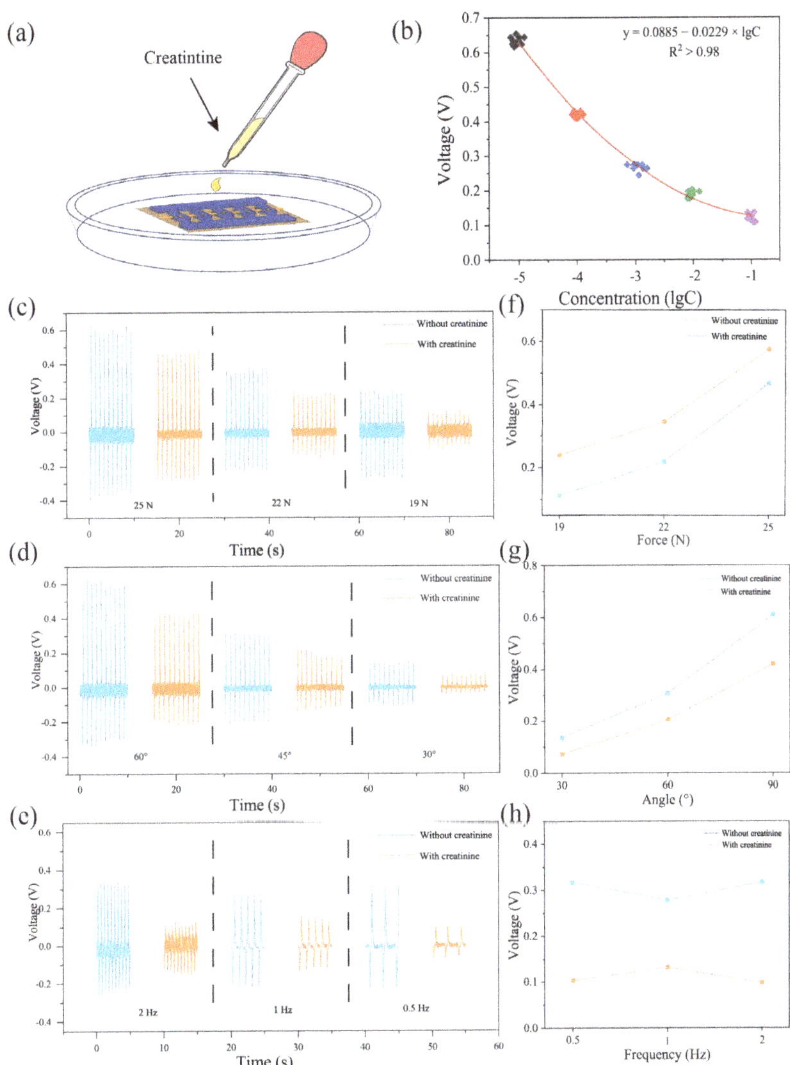

Figure 4. (**a**) Measurement process. (**b**) Relationship between creatinine concentration and piezoelectric output voltage. Sensing performance against different applied forces (**c**), bending angles (**d**), and force frequencies (**e**). Comparison of piezoelectric output variation trends under different force sizes (**f**), bending angle (**g**), and force frequencies (**h**) before and after adding creatinine.

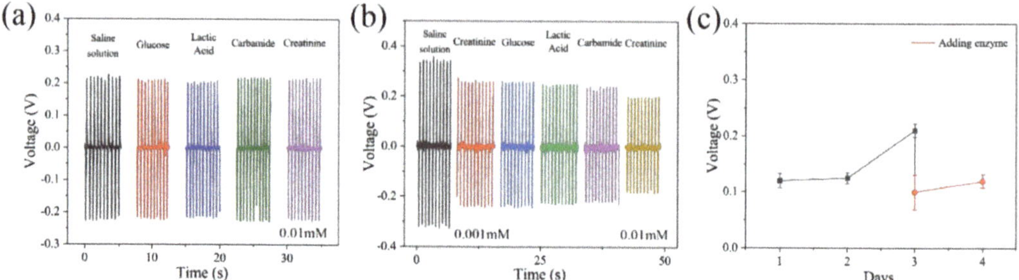

Figure 5. The piezoelectric output voltage of the device before (**a**) and after (**b**) modifying enzymes. (**c**) Output of the device across creatinine solutions for the first four days.

Figure 5 shows the selective specificity of the self-powered creatinine biosensor. The concentration of all solutions used in the experiment is 0.01 mM. Figure 5a shows the piezoelectric output of the device without modifying enzymes. In this experiment, the piezoelectric output voltage of the device without modifying enzymes is similar against saline, glucose, lactic acid, urea, and creatinine. However, as shown in Figure 5b, after adding creatinine (0.001 mM), the piezoelectric output voltage decreases (red line). The addition of other materials (glucose, lactic acid, and carbamide) (0.001 mM) has little effect on the piezoelectric output voltage. Finally, increasing the concentration of the creatinine solution to 0.01 mM lowers the piezoelectric output voltage (yellow line). These findings suggest that enzymatic reactions can influence the piezoelectric output voltage of the ZnO NWs and that creatininase/creatinase/sarcosine oxidase-modified ZnO NWs have the potential to be used in the fabrication of self-powered biosensors. Figure 5c shows the output of the device across creatinine solutions (0.1 mM) for the first four days. It can be seen that piezoelectric output voltage increases on the third day. However, by timely replenishing enzymes, the voltage returns to its initial state.

Table 1 is a comparison of the performance parameters of two common commercial creatinine sensors and the self-powered creatinine biosensor. As shown in the table, the limit of detection parameter of the self-powered creatinine biosensor is lower than that of the two commercial creatinine detectors. Due to its small size and lack of external power supply, the self-powered creatinine biosensor is superior in terms of portability. Therefore, the self-powered creatinine sensor has a good application prospect for daily life.

Table 1. Comparison of performance of three creatinine detectors.

	Limit of Detection	Limit of Quantification	Size	Power Supply	Portability
ACON U120smart	0.08 mM	0.08–2.56 mM	27 × 18 × 14 cm	External power	Not portable
On Call CMU060	0.08 mM	0.08–2.56 mM	14 × 13 × 4 cm	External power	Not portable
This work	1×10^{-5} mM	1×10^{-5}–1×10^{-1} mM	2 × 3 × 0.2 mm	Self-powered	Portable

Figure 6 illustrates the mechanism of the piezo-enzymatic-reaction effect. The simulation results are from COMSOL Multiphysics 5.5. When no force is applied, the piezoelectric output voltage is 0 V, as shown in Figure 6a. When the ZnO NWs are deformed by an applied force, the piezoelectric output voltage is proportional to the c-axis external strain (Figure 6b) [42]. The larger force and higher angles can induce a larger deformation, which improves the output (Figure 3b,c). The enzymatic reaction is shown in Figure 6(c-i) [43]. After the reaction, H_2O_2 is produced as:

$$Creatinine + H_2O + O_2 \xrightarrow{Creatininase/Creatinase/Arcosine\ Oxidase} Formaldehyde + Glycine + H_2O_2, \quad (4)$$

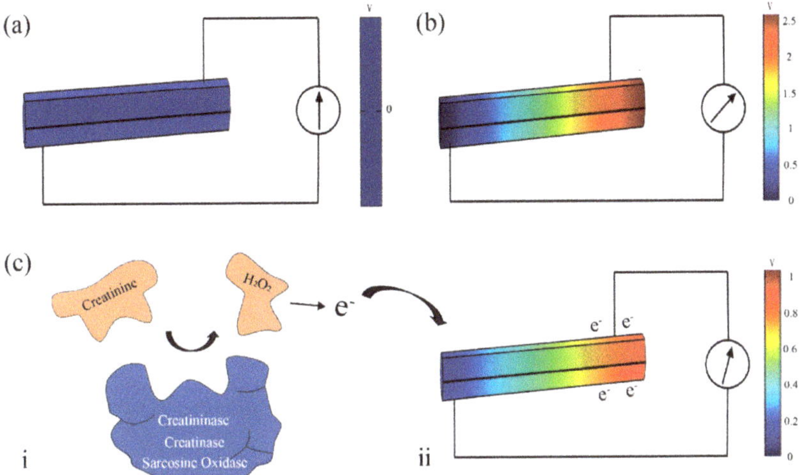

Figure 6. (**a**,**b**) COMSOL simulation of piezoelectric effect. (**c**) Enzymatic reactions and coupling effects.

H_2O_2 is unstable and H^+ and e^- are produced [36] as:

$$H_2O_2 \rightarrow 2e^- + 2H^+ + O_2 \quad (5)$$

The large number of H^+ and e^- have directional movement and screen the built-in electric field of the ZnO NWs under deformation (Figure 6(c-ii)). The higher concentration of creatinine solution can release more H^+ and e^-, which enhance the screen effect and further lower the piezoelectric output voltage. Other materials are weak electrolytes that are difficult to ionize to enhance the screen effect. Therefore, when the enzymatic reaction occurs, the piezoelectric output voltage will decrease.

4. Conclusions

In summary, a self-powered creatinine biosensor was built. The device can detect creatinine concentration in sweat in real time using the piezo-enzymatic-reaction effect of ZnO NWs, and the entire sensing process requires no external power supply. The detection range of creatinine concentration of the self-powered creatinine biosensor was 1×10^{-5}–1×10^{-1} mM and the sensitivity was 0.0229 V/mM. This result indicates that the creatinine sensor has a wide application prospect and has great significance for the prevention of related diseases.

Author Contributions: Conceptualization, T.Z. and Q.W.; methodology, T.Z.; software, G.Z.; validation, T.Z., M.W., and G.Z.; formal analysis, T.Z.; investigation, T.Z.; resources, T.Z.; data curation, M.W.; writing—original draft preparation, M.W.; writing—review and editing, T.Z. and Q.W.; visualization, M.W., J.L., Y.S. and X.Z.; supervision, Q.W.; project administration, Q.W.; funding acquisition, Q.W. All authors have read and agreed to the published version of the manuscript.

Funding: This research was funded by Fundamental Research Funds for the Central Universities, grant number N2105011.

Institutional Review Board Statement: Not applicable.

Informed Consent Statement: Not applicable.

Data Availability Statement: Data is contained within the article.

Acknowledgments: Thanks to Jilong Gao from Shiyanjia Lab (www.shiyanjia.com (accessed on 14 August 2021)) for the XRD analysis and Duan from Shiyanjia Lab (www.shiyanjia.com (accessed on 9 September 2021)) for the paper polishing service.

Conflicts of Interest: The authors declare no conflict of interest.

References

1. Choe, W.; Chae, J.D.; Lee, B.H.; Kim, S.H.; Park, S.Y.; Nimse, S.B.; Kim, J.; Warkad, S.D.; Song, K.S.; Oh, A.C.; et al. 9G Test (TM) Cancer/Lung: A Desirable Companion to LDCT for Lung Cancer Screening. *Cancers* **2020**, *12*, 3192. [CrossRef] [PubMed]
2. Song, K.S.; Nimse, S.B.; Warkad, S.D.; Oh, A.C.; Kim, T.; Hong, Y.J. Quantification of CYFRA 21-1 and a CYFRA 21-1-anti-CYFRA 21-1 autoantibody immune complex for detection of early stage lung cancer. *Chem. Commun.* **2019**, *55*, 10984. [CrossRef] [PubMed]
3. Park, M.; Heo, Y.J. Biosensing Technologies for Chronic Diseases. *Biochip J.* **2021**, *15*, 1–13. [CrossRef]
4. Kim, J.Y.; Sung, G.Y.; Park, M. Efficient Portable Urea Biosensor Based on Urease Immobilized Membrane for Monitoring of Physiological Fluids. *Biomedicines* **2020**, *8*, 596. [CrossRef] [PubMed]
5. Gao, D.H.; Yang, X.H.; Teng, P.P.; Luo, M.; Zhang, H.X.; Liu, Z.H.; Yang, J.; Li, Z.N.; Wen, X.Y.; Yuan, L.B.; et al. In-fiber optofluidic online SERS detection of trace uremia toxin. *Opt. Lett.* **2021**, *46*, 1101–1104. [CrossRef]
6. Pundir, C.S.; Yadav, S.; Kumar, A. Creatinine sensors. *Trends Analyt. Chem.* **2013**, *50*, 42–52. [CrossRef]
7. Lad, U.; Khokhar, S.; Kale, G.M. Electrochemical Creatinine Biosensors. *Anal. Chem.* **2008**, *80*, 7910–7917. [CrossRef]
8. Wyss, M.; Kaddurah-Daouk, R. Creatine and creatinine metabolism. *Physiol. Rev.* **2000**, *80*, 1107–1213. [CrossRef]
9. Killard, A.J.; Smyth, M.R. Creatinine biosensors: Principles and designs. *Trends Biotechnol.* **2000**, *18*, 433–437. [CrossRef]
10. Panasyuk-Delaney, T.; Mirsky, V.M.; Wolfbeis, O.S. Capacitive creatinine sensor based on a photografted molecularly imprinted polymer. *Electroanalysis* **2002**, *14*, 221–224. [CrossRef]
11. Cholongitas, E.; Marelli, L.; Kerry, A.; Senzolo, M.; Goodier, D.W.; Nair, D.; Thomas, M.; Patch, D.; Burroughs, A.K. Different methods of creatinine measurement significantly affect MELD scores. *Liver* **2007**, *13*, 523–529. [CrossRef]
12. Narayanan, S.; Appletion, H.D. Creatinine—A Review. *Clin. Chem.* **1980**, *26*, 1119–1126. [CrossRef]
13. Cirillo, M.; Laurenzi, M.; Mancini, M.; Zanchetti, A.; De Santo, N.G. Low muscular mass and overestimation of microalbuminuria by urinary albumin/creatinine ratio. *Hypertension* **2003**, *47*, 56–61. [CrossRef]
14. Luo, Y.; Zhao, T.M.; Dai, Y.T.; Li, Q.; Fu, H.Y. Flexible nanosensors for non-invasive creatinine detection based on triboelectric nanogenerator and enzymatic reaction. *Sens. Actuators A Phys.* **2021**, *320*, 112585. [CrossRef]
15. Guan, H.Y.; Lv, D.; Zhong, T.Y.; Dai, Y.T.; Xing, L.L.; Xue, X.Y.; Zhang, Y.; Zhan, Y. Self-powered, wireless-control, neural-stimulating electronic skin for in vivo characterization of synaptic plasticity. *Nano Energy* **2020**, *67*, 104182. [CrossRef]
16. He, H.X.; Zhao, T.M.; Guan, H.Y.; Zhong, T.Y.; Zeng, H.; Xing, L.L.; Zhang, Y.; Xue, X.Y. A water-evaporation-induced self-charging hybrid power unit for application in the Internet of Things. *Sci. Bull.* **2019**, *64*, 1409–1417. [CrossRef]
17. Dong, T.; Peng, P.; Jiang, F.M. Numerical modeling and analysis of the thermal behavior of NCM lithium-ion batteries subjected to very high C-rate discharge/charge operations. *Int. J. Heat Mass Transf.* **2018**, *117*, 261–272. [CrossRef]
18. Sun, K.; Wei, T.S.; Ahn, B.Y.; Seo, J.Y.; Dillon, S.J.; Lewis, J.A. 3D Printing of Interdigitated Li-Ion Microbattery Architectures. *Adv. Mater.* **2013**, *25*, 4539–4543. [CrossRef] [PubMed]
19. Xu, M.Y.; Wang, P.H.; Wang, Y.C.; Zhang, S.L.; Wang, A.C.; Zhang, C.L.; Wang, Z.J.; Pan, X.X.; Wang, Z.L. Soft and Robust Spring Based Triboelectric Nanogenerator for Harvesting Arbitrary Directional Vibration Energy and Self-Powered Vibration Sensing. *Adv. Energy Mater.* **2018**, *8*, 1702432. [CrossRef]
20. Surmenev, R.A.; Orlova, T.; Chernozem, R.V.; Ivanova, A.A.; Bartasyte, A.; Mathur, S.; Surmeneva, M.A. Hybrid lead-free polymer-based nanocomposites with improved piezoelectric response for biomedical energy-harvesting applications: A review. *Nano Energy* **2019**, *62*, 475–506. [CrossRef]
21. Shi, Q.F.; Wu, H.; Wang, H.; Wu, H.X.; Lee, C.K. Self-Powered Gyroscope Ball Using a Triboelectric Mechanism. *Adv. Energy Mater.* **2017**, *7*, 1701300. [CrossRef]
22. Guo, H.Y.; Wen, Z.; Zi, Y.L.; Yeh, M.H.; Wang, J.; Zhu, L.P.; Hu, C.G.; Wang, Z.L. A Water-Proof Triboelectric-Electromagnetic Hybrid Generator for Energy Harvesting in Harsh Environments. *Adv. Energy Mater.* **2015**, *6*, 1501593. [CrossRef]
23. Lee, H.; Kim, H.; Kim, D.Y.; Seo, Y. Pure Piezoelectricity Generation by a Flexible Nanogenerator Based on Lead Zirconate Titanate Nanofibers. *ACS Omega* **2019**, *4*, 2610–2617. [CrossRef]
24. Jin, C.R.; Hao, N.J.; Xu, Z.; Trase, I.; Nie, Y.; Dong, L.; Closson, A.; Chen, Z.; Zhang, J.X.J. Flexible piezoelectric nanogenerators using metal-doped ZnO-PVDF films. *Sens. Actuators A Phys.* **2020**, *305*, 111912. [CrossRef] [PubMed]
25. Tang, X.Y.; Wang, J.C.; Liu, X.J.; Yang, L.; Liu, B.D.; Jiang, X. Fabrication of $CuO/TiO_2/Ti$ monolithic catalyst for efficient and stable CO oxidation. *Adv. Mater. Interfaces* **2021**, *8*, 21004400. [CrossRef]
26. Kim, M.P.; Um, D.S.; Shin, Y.E.; Ko, H. High-Performance Triboelectric Devices via Dielectric Polarization: A Review. *Nanoscale Res. Lett.* **2021**, *16*, 35. [CrossRef]
27. Wang, R.C.; Lin, Y.C.; Chen, P.T.; Chen, H.C.; Chiu, W.T. Anomalous output performance enhancement of RGO-based triboelectric nanogenerators by Cu-bonding. *Nano Energy* **2021**, *86*, 106126. [CrossRef]
28. Li, S.Y.; Fan, Y.; Chen, H.Q.; Nie, J.H.; Liang, Y.X.; Tao, X.L.; Zhang, J.; Chen, X.Y.; Fu, E.G.; Wang, Z.L. Manipulating the triboelectric surface charge density of polymers by low-energy helium ion irradiation/implantation. *Energy Environ. Sci.* **2020**, *13*, 896–907. [CrossRef]

29. Li, J.; Zhang, X.L.; Yang, B.; Zhang, C.; Xu, T.T.; Chen, L.X.; Yang, L.; Jin, X.; Liu, B.D. Template approach large-layered non-layer Ga-group two-dimensional crystals from printed skin of liquid gallium. *Mater. Chem. Phys.* **2021**, *33*, 4568–4577.
30. Liu, N.J.; Chu, L.; Ahmad, W.; Hu, R.Y.; Luan, R.F.; Liu, W.; Yang, J.; Ma, Y.H.; Li, X.A. Low-pressure treatment of CuSCN hole transport layers for enhanced carbon-based perovskite solar cells. *J. Power Sources* **2021**, *499*, 229970. [CrossRef]
31. Ma, Y.H.; Zhang, Y.W.; Zhang, H.Y.; Lv, H.; Hu, R.Y.; Liu, W.; Wang, S.L.; Jiang, M.; Chu, L.; Zhang, J.; et al. Effective carrier transport tuning of CuOx quantum dots hole interfacial layer for high-performance inverted perovskite solar cell. *Appl. Surf. Sci.* **2021**, *547*, 149117. [CrossRef]
32. Xiang, C.H.; Liu, C.R.; Hao, C.L.; Wang, Z.K.; Che, L.F.; Zhou, X.F. A self-powered acceleration sensor with flexible materials based on triboelectric effect. *Nano Energy* **2017**, *31*, 469–477. [CrossRef]
33. Luo, X.X.; Zhu, L.P.; Wang, Y.C.; Li, J.Y.; Nie, J.J.; Wang, Z.L. A Flexible Multifunctional Triboelectric Nanogenerator Based on MXene/PVA Hydrogel. *Adv. Funct. Mater.* **2021**, 2104928. [CrossRef]
34. Ashiba, H.; Oyamada, C.; Hosokawa, K.; Ueno, K.; Fujimaki, M. Sensitive Detection of C-Reactive Protein by One-Step Method Based on a Waveguide-Mode Sensor. *Biosensors* **2020**, *20*, 3258. [CrossRef]
35. Mao, Y.P.; Zhu, Y.S.; Zhao, T.M.; Jia, C.J.; Bian, M.Y.; Li, X.X.; Liu, Y.G.; Liu, B.D. A Portable and Flexible Self-Powered Multifunctional Sensor for Real-Time Monitoring in Swimming. *Biosensors* **2021**, *11*, 147. [CrossRef] [PubMed]
36. Xue, X.Y.; Qu, Z.; Fu, Y.M.; Yu, B.W.; Xing, L.L.; Zhang, Y. Self-powered electronic-skin for detecting glucose level in body fluid basing on piezo-enzymatic-reaction coupling process. *Nano Energy* **2016**, *26*, 148–156. [CrossRef]
37. Zhao, T.M.; Fu, Y.M.; He, H.X.; Dong, C.Y.; Zhang, L.L.; Zeng, H.; Xing, L.L.; Xue, X.Y. Self-powered gustation electronic skin for mimicking taste buds based on piezoelectric-enzymatic reaction coupling process. *Nanotechnology* **2018**, *29*, 075501. [CrossRef] [PubMed]
38. Mao, Y.P.; Shen, M.L.; Liu, B.; Xing, L.L.; Chen, S.; Xue, X.Y. Self-Powered Piezoelectric-Biosensing Textiles for the Physiological Monitoring and Time-Motion Analysis of Individual Sports. *Sensors* **2019**, *19*, 3310. [CrossRef]
39. Patel, S.S.; Molnar, M.Z.; Tayek, J.A.; Ix, J.H.; Noori, N.; Benner, D.; Heymsfield, S.; Kopple, J.D.; Kovesdy, C.P.; Kalantar-Zadeh, K. Serum creatinine as a marker of muscle mass in chronic kidney disease: Results of a cross-sectional study and review of literature. *J. Cachexia Sarcopenia Muscle* **2013**, *4*, 19–29. [CrossRef] [PubMed]
40. Colombo, M.; McGurnaghan, S.J.; Blackbourn, L.A.K.; Dalton, R.N.; Dunger, D.; Bell, S.; Petrie, J.R.; Green, F.; MacRury, S.; McKnight, J.A.; et al. Comparison of serum and urinary biomarker panels with albumin/creatinine ratio in the prediction of renal function decline in type 1 diabetes. *Diabetologia* **2020**, *63*, 788–798. [CrossRef] [PubMed]
41. Han, W.X.; He, H.X.; Zhang, L.L.; Dong, C.Y.; Zeng, H.; Dai, Y.T.; Xing, L.L.; Zhang, Y.; Xue, X.Y. A Self-Powered Wearable Noninvasive Electronic-Skin for Perspiration Analysis Based on Piezo-Biosensing Unit Matrix of Enzyme/ZnO Nanoarrays. *ACS Appl. Mater. Interfaces* **2017**, *9*, 29526–29537. [CrossRef] [PubMed]
42. Zhao, T.M.; Wang, Q.; Du, A. High piezo-photocatalytic efficiency of H_2 production by CuS/ZnO. *Mater. Lett.* **2021**, *294*, 129752. [CrossRef]
43. Vendruscolo, M.; Dobson, C.M. Dynamic visions of enzymatic reactions. *Science* **2006**, *313*, 1586–1587. [CrossRef] [PubMed]

MDPI
St. Alban-Anlage 66
4052 Basel
Switzerland
Tel. +41 61 683 77 34
Fax +41 61 302 89 18
www.mdpi.com

Biosensors Editorial Office
E-mail: biosensors@mdpi.com
www.mdpi.com/journal/biosensors

www.ingramcontent.com/pod-product-compliance
Lightning Source LLC
LaVergne TN
LVHW070626170726
843515LV00004B/192

* 9 7 8 3 0 3 6 5 3 5 6 9 2 *